Exiled to Palestine

This book tells the largely unknown story of how Zionists imprisoned by the Soviet authorities in the 1920s and 1930s were permitted to opt for a sentence of permanent exile to Palestine. There, they made a significant contribution to building a Jewish polity – forming the backbone of influential left-wing parties and the powerful trade union movement.

Utilizing fresh documents from archives opened after the collapse of the Soviet Union, as well as British and Zionist sources, the authors examine the means by which internal power struggles and personal interventions in the uppermost echelons of the Soviet leadership enabled the Zionists to disseminate their message and recruit thousands of members before the massive arrests of the mid-1920s. They further reveal the extent to which personal contacts between Zionists and Soviet officials were vital in initiating and sustaining the phenomenon of exile to Palestine and assess the crucial role of Anglo-Soviet cooperation in facilitating the immigration of Zionist convicts.

A selection of twenty-two translated and annotated documents from Israeli and Russian archival collections is included.

This book will be of great interest to all students of Jewish and Israeli history, Russian and Soviet studies and the history of British rule in Palestine.

Ziva Galili is Professor of Russian and Soviet History and Chair of the History Department at Rutgers University in New Jersey. She is an authority on the Menshevik Party and is currently at work on a study of Zionism in Soviet Russia in the 1920s.

Boris Morozov is a Research Fellow at the Cummings Center for Russian and East European Studies. From 1978 until 1984 he worked at the Institute for Documental Research and Archives of the Central Soviet Archives (Glavarkhiv SSSR) in Moscow. He specializes in the methodology of archival research and is the author of *Documents on Soviet Jewish Emigration* (1999).

The Cummings Center for Russian and East European Studies

 *Tel Aviv University*

The Cummings Center is Tel Aviv University's main framework for research, study, documentation and publication relating to the history and current affairs of Russia, the former Soviet republics and Eastern Europe. The Center is committed to pursuing projects which make use of fresh archival sources and to promoting a dialogue with Russian academic circles through joint research, seminars and publications.

The Cummings Center Series

The titles published in this series are the product of original research by the Center's faculty, research staff and associated fellows. The Cummings Center Series also serves as a forum for publishing declassified Russian archival material of interest to scholars in the fields of history and political science.

Editor-in-Chief
Gabriel Gorodetsky

Editorial Board
Michael Confino
Igal Halfin
Yaacov Ro'i
Nurit Schleifman

Managing Editor
Deena Leventer

ISSN 1365-3733

Titles in the Cummings Center Series

1. **Soviet Foreign Policy 1917–1991**
   A retrospective
   Edited by Gabriel Gorodetsky

2. **Jews and Jewish Life in Russia and the Soviet Union**
   Edited by Yaacov Ro'i

3. **Muslim Eurasia**
   Conflicting legacies
   Edited by Yaacov Ro'i

4. **Envoy to Moscow**
   Memoirs of an Israeli Ambassador, 1988–92
   Aryeh Levin

5. **Egypt's Incomplete Revolution**
   Lutfi al-Khuli and Nasser's Socialism in the 1960s
   Rami Ginat

6. **Russian Jews on Three Continents**
   Migration and resettlement
   Edited by Noah Lewin-Epstein, Yaacov Ro'i and Paul Ritterband

7. **In Pursuit of Military Excellence**
   The evolution of operational theory
   Shimon Naveh

8. **Russia at a Crossroads**
   History, memory and political practice
   Edited by Nurit Schleifman

9. **Political Organization in Central Asia and Azerbaijan**
   Sources and documents
   Edited by Vladimir Babak, Demian Vaisman and Aryeh Wasserman

11. **The 1956 War**
    Collusion and rivalry in the Middle East
    Edited by David Tal

13. **Documents on Israeli–Soviet relations, 1941–1953**
    Part I: 1941–May 1949;
    Part II: May 1949–1953

14. **Documents on Soviet Jewish Emigration**
    Boris Morozov

15. **Documents on Ukrainian Jewish Identity and Emigration, 1944–1990**
    Vladimir Khanin

16. **Language and Revolution**
    Making modern political identities
    Edited by Igal Halfin

17. **Stalin and the Inevitable War 1936–1941**
    Silvio Pons

18. **Russia between East and West**
    Russian foreign policy on the threshold of the twenty-first century
    Edited by Gabriel Gorodetsky

19. **Democracy and Pluralism in Muslim Eurasia**
    Edited by Yaacov Ro'i

20. **Educational Reform in Post-Soviet Russia**
    Legacies and prospects
    Edited by Ben Eklof, Larry E. Holmes and Vera Kaplan

21. **Exiled to Palestine**
    The emigration of Zionist convicts from the Soviet Union, 1924–1934
    Ziva Galili and Boris Morozov

Exiled to Palestine

The emigration of Zionist convicts from the Soviet Union, 1924–1934

Ziva Galili and Boris Morozov

Routledge
Taylor & Francis Group
LONDON AND NEW YORK

First published 2006
by Routledge
2 Park Square, Milton Park, Abingdon, Oxon OX14 4RN

Simultaneously published in the USA and Canada
by Routledge
270 Madison Ave, New York, NY 10016

Routledge is an imprint of the Taylor & Francis Group

© 2006 Ziva Galili and Boris Morozov

Typeset in Times New Roman by
Newgen Imaging Systems (P) Ltd, Chennai, India
Printed and bound in Great Britain by
Biddles Ltd, King's Lynn

All rights reserved. No part of this book may be reprinted or
reproduced or utilised in any form or by any electronic,
mechanical, or other means, now known or hereafter
invented, including photocopying and recording, or in any
information storage or retrieval system, without permission in
writing from the publishers.

The publisher makes no representation, express or implied,
with regard to the accuracy of the information contained in
this book and cannot accept any legal responsibility or liability
for any errors or omissions that may be made.

British Library Cataloguing in Publication Data
A catalogue record for this book is available from the British Library

Library of Congress Cataloging in Publication Data
Galili y Garcia, Ziva.
  Exiled to Palestine : the emigration of Soviet Zionist convicts,
1924-1934 / Ziva Galili, Boris Morozov.
     p. cm. – (Cummings Center series)
  Includes index.
     1. Jews–Soviet Union–Migrations–History–20th century.
  2. Zionism–Soviet Union–History. 3. Antisemitism–Soviet
  Union–History. 4. Political refugees–Israel–History. 5. Jews,
  Soviet–Israel–History. 6. Soviet Union–Emigration and immigration.
  7. Soviet Union–Ethnic relations. I. Morozov, Boris. II. Title. III. Series.
  DS135.R92G35 2006
  304.8′569404709042–dc22   2005028473

ISBN10: 0–714–65708–5
ISBN13: 9–78–0–714–65708–0

Contents

Preface		viii
Acknowledgements		x
List of documents		xi
1	**Introduction: Zionism in Soviet Russia** ZIVA GALILI	1
2	**Out of the Soviet Union: the exiles and Pompolit** BORIS MOROZOV	14
3	**Into Palestine: the Zionists and the British** ZIVA GALILI	41
4	**Postscript: two fates** ZIVA GALILI	75
5	**Documents**	85
	Glossary	134
	Index	137

Preface

The co-authors of this volume each arrived at the study of 'substitution immigration' through a personal discovery. Working in the 1980s at the Central Archive of the October Revolution in Moscow (later renamed the State Archive of the Russian Federation), Boris Morozov uncovered massive evidence of the help rendered by the Committee to Aid Political Prisoners during the 1920s and 1930s to Zionist prisoners seeking emigration from Soviet Russia to Palestine. He read and sorted through hundreds of documents and assembled data on numerous Zionists aided by the Committee. Ziva Galili became familiar with substitution immigration through the stories of her mother and other members of her native Kibbutz Afikim in Israel, who owed their passage from prison to Palestine to this unique arrangement. But it was only in the late 1990s, when she began to study Zionism in Soviet Russia and the role of its members in shaping Jewish society in Palestine, that she discovered how many of the pioneers who arrived during the decade from 1924 through 1934 had benefited from the substitution arrangement. Her search for clues to the origins of this phenomenon led to archives in Israel, Britain and Russia.

The two projects were brought together by Gabriel Gorodetsky, Director of the Cummings Center and General Editor of its Series, who has been a source of encouragement and editorial advice throughout. Expert opinion and helpful critique came also from Yaacov Ro'i. The authors wish to acknowledge their enormous debt to the Cummings Series Editor, Deena Leventer, who spared no effort in seeing the volume to its conclusion. Without her sharp eye and wise counsel this work would have been much the poorer.

This small yet significant chapter remains unknown in the history of Soviet Russia, while the few published histories of Russian Zionism merely record its existence.[1] Contemporaries knew about it from newspapers and journals published in Palestine; veterans of Soviet Zionist movements recorded it in their memoirs. However, the information available in print is based largely on the personal experiences of those granted substitution and does not explain how the arrangements developed and operated over time. In order to gain a deeper and more complete understanding of the phenomenon, the authors made extensive use of archives in Russia, Israel and England – primarily of those organizations engaged in setting policy and implementing the substitution and the immigrants' passage to Palestine.

Documents concerning the origins of substitution and the considerations that guided the Soviet government are still rare, but several have been used here: individual documents of the United State Political Directorate found in the papers of Feliks Dzerzhinskii[2] and the Central Committee of the Communist Party,[3] documents of the Central Committee's Jewish Section (Evsektsiia),[4] and the diary kept by David Shor,[5] a renowned Russian-Jewish pianist who served as the linchpin in organizing meetings between Zionist representatives and Soviet leaders. In contrast, the implementation of substitution in Soviet Russia is documented massively, at least from the second half of 1925. It was at that time that the Committee to Aid Political Prisoners in Moscow took over coordination of all aspects of the process in Soviet Russia, leaving behind hundreds of files of correspondence with Zionist exiles, their families, Soviet authorities and the Immigration Centre (Merkaz ha-Aliya) of the Federation of Jewish Workers in Palestine (Histadrut).[6] These sources are employed extensively in the chapter entitled, Out of the Soviet Union.

Chapter 3, Into Palestine, draws upon on several archival collections in Israel and Britain to examine the politics of substitution immigration into Palestine. The archive of the British mandatory Government of Palestine[7] and the papers of the Colonial and Foreign offices in London are used to document British policy and practice.[8] The collections of the Political Department of the World Zionist Executive in London and the Immigration and Political Departments of the Palestine Zionist Executive in Jerusalem shed light on the role played by these organizations.[9] The role of the Federation of Jewish Workers in Palestine can be appreciated from the extant papers of its Immigration Centre.[10] Finally, documents produced by representatives in Palestine of the Soviet Zionist movements testify to their significant share in making the substitution immigration possible. Split among several collections and archives in Israel, these documents contain priceless information for a history, still to be written, of Zionism in Soviet Russia.[11]

Notes

1 See, for example, Yitzchak Ma'or, *Ha-Tnu'a ha-Tziyonit be-Rusya me-Reshita ve-ad Yameinu* (Jerusalem: Ha-Sifriya ha-Tziyonit & Magnes, 1973), pp. 538–9.
2 Rossiiskii Gosudarstvennyi Arkhiv Sotsial'noi i Politicheskoi Istorii (RGASPI), Moscow, Fond 76.
3 RGASPI, Fond 17.
4 RGASPI, Fond 445.
5 Manuscript Collection, National and University Libraries (NUL), Jerusalem.
6 Gosudarstvennyi Arkhiv Rossiiskoi Federatsii (GARF), Moscow, Fond 8049.
7 Israel State Archives (ISA), Jerusalem.
8 Public Record Office (PRO), London.
9 Central Zionist Archives (CZA), Jerusalem.
10 Lavon Institute for Labour Research, Tel Aviv (IV-211).
11 The Arye Zenziper (Tzentziper) Collection (F30) at the CZA; Binyamin Vest collection (IV-104-53) at the Lavon Archive; papers of the socialist wing of Ha-Shomer ha-Tza'ir at the Archive of Kibbutz Afikim; documents of other socialist movements at the Archive of Ha-Kibbutz ha-Me'uchad, Yad Tabenkin at Ramat Ef'al.

Acknowledgements

The publication of this book was made possible through a generous donation of the Agmon family, in memory of the late Avraham Agmon. The Agmon project was established in order to study and document the struggle of Soviet Jews for emigration to Israel. The Cummings Center is also grateful to the Lucius N. Littauer Foundation of New York for its support of this research.

The resources and capable staff of The Central Zionist Archive, Israel State Archives, The Lavon Institute for Labour Research, Yad Tabenkin Archive, the Beth Hatefutsoth Visual Documentation Center, the State Archive of the Russian Federation (GARF) and the Russian State Archive for Contemporary Political History (RGASPI) were vital to the writing of this work. The authors would especially like to thank Kirill M. Andersen, Sergei V. Mironenko, Ol'ga L. Milova, Rochelle Rubinstein, Alexandra Tumarinson and Zippi Rosenne for their expert advice and assistance.

Documents

1. **Mandate given to E.P. Peshkova by the Deputy Chairman of the GPU**
   Russian
   GARF f. 8409, op. 1, d. 11, l. 1
   11 November 1922

2. **Memorandum of A. Merezhin, Central Bureau of the Evsektsiia to the Politburo**
   Russian
   RGASPI f. 445, op. 1, d. 119, ll. 120–1
   March 1923

3. **Report of the OGPU on the search and arrest of Zionist Activists**
   Russian
   RGASPI f. 17, op. 84, d. 643, ll. 4–6
   March 1924

4. **F.E. Dzerzhinskii to his Deputies V.R. Menzhinskii and G.G. Iagoda**
   Russian
   RGASPI f. 76, op. 3, d. 326, ll. 2–3
   15 March 1924

5. **Y. Mereminsky, Immigration Centre to Central Office of World Hechalutz (Berlin)**
   Hebrew
   Lavon Archive IV-211-3, f. 2
   22 October 1924

6. **Y. Mereminsky, Immigration Centre to Palestine Zionist Executive**
   English
   Lavon Archive IV-211-3, f. 17A
   10 December 1924

7. **F.E. Dzerzhinskii to V.R. Menzhinskii**
   Russian
   RGASPI f. 76, op. 3, d. 326, l. 4
   24 March 1925

xii *Documents*

8 Report of Special Department of OGPU to F.E. Dzerzhinskii
Russian
RGASPI f. 76, op. 3, d. 326, l. 5
29 May 1925

9 A.M. Hyamson, Chief Immigration Officer, Government of Palestine to British Chargé d'Affaires (Moscow)
English
ISA 11/1147, CONS/E/37/2
16 September 1925

10 Central Eretz Yisrael Office (Warsaw) to Immigration Department, Palestine Zionist Executive (Jerusalem)
Hebrew
Copy: CZA S25/f. 2424
12 April 1927

11 A.M. Hyamson, Chief Immigration Officer, Government of Palestine to F.H. Kisch, Chairman of PZE
English
ISA 11/1174, IMM/7 I
30 June 1927

12 Memorandum of Norwegian Legation on the procedure for obtaining Palestine visas for Soviet citizens
English
ISA 11/1174, IMM/7 I
18 July 1927

13 F.H. Kisch, Chairman of PZE to M. Nurock, Assistant Chief Secretary of the Government of Palestine
English
CZA S25/f. 2424
26 July 1927

14 H.O. Plumer, High Commissioner for Palestine to L. Amery, Secretary of State for the Colonies
English
ISA 11/1174, IMM/7 I
5 October 1927

15 Correspondence concerning the substitution of Y.S. Arav and A.L. Molodetskaia-Arav (September 1926–October 1927)
Russian
GARF f. 8409, op. 1, d. 1698, l. 60
24 September 1926

16 Ch. Halperin, Immigration Centre to Pompolit
Russian
GARF f. 8409, op. 1, d. 232, ll. 34, 234-ob.
14 June 1928

17 **Pompolit to Immigration Centre**
    Russian
    GARF f. 8409, op. 1, d. 232
    12 December 1928

18 **Central Office of the Zionist Organization (London) to members of the Executive Committee and to the Zionist federations and fractions**
    Hebrew
    Lavon Archive IV-104-53, f. 35
    4 March 1929

19 **A. Saunders, Department of Police and Prisons, Government of Palestine to Chief Immigration Officer**
    English
    ISA 11/1174, IMM/7 II
    12 April 1929

20 **Palestine Zionist Executive to the Chief Secretary, Government of Palestine**
    English
    CZA S6/f. 5355
    10 June 1929

21 **Draft report of Interdepartmental Committee, Government of Palestine on immigration from Russia**
    English
    ISA 11/1174/2335, IMM/7/8
    October 1930

22 **Y. Gruenbaum, Jewish agency to Director of Immigration, Government of Palestine**
    English
    ISA 11/1174, IMM/7/6 II
    26 February 1934

1 Introduction
Zionism in Soviet Russia

Ziva Galili

> At the present time we face quite a serious situation: our means of administrative struggle against the Zionist movement are not attaining their goal since active Zionist forces are arising with frightening alacrity from the depths of the Jewish masses and the predominant majority of these forces are the youth....
> It would be no exaggeration to say that there is not a single Jewishly-inhabited location of any size in Ukraine without a Zionist cell or group that is active in all spheres of life of that location and even prevails in terms of its influence and leadership role among local masses over Communist cells and public or government organizations.[1]
>
> (From a secret memorandum by Eduard Karlson, deputy chairman of the Ukrainian State Political Directorate to Secretary of the Central Committee, CP Ukraine, Lazar' Kaganovich, 1925)

The proliferation of Zionist activity described in this internal Soviet document lies at the base of a unique and intriguing episode in Soviet history involving the exit of over a thousand Zionists from Soviet Russia during the decade from 1924 to 1934. Jews, arrested for carrying out Zionist activity, were permitted by the Soviet authorities to opt for 'deportation' to Palestine in place of prison or exile to remote reaches of the Soviet imperium. The 'substitution' emigration allowed a small but committed minority from among the many thousands of active Zionists in Soviet Russia to reach the shores of Palestine, where they participated in the building of a new Jewish society and polity. Their story is significant, both for the early history of Soviet rule in Russia, and for the shaping of Israeli society and culture. Yet, it has never been told or documented beyond the memoirs and oral legends passed down by veteran Zionists as they looked back with wonder and some sorrow at the 'miracle' that saved them but left many of their friends and comrades in Soviet Russia.

This volume aims to fill the void and reconstruct, based on Israeli, British and Soviet sources, an account of how the escape route was opened by the Soviet authorities and turned into a channel for pioneering immigration into Palestine. On the Soviet side, this was made possible through the policies towards nationalities in general and the Jewish question in particular, as well as through the political considerations born out of the struggle within the Soviet leadership, and the tireless efforts of Ekaterina Peshkova, the indomitable head of the Committee to Aid Political Prisoners (Pompolit). Outside Soviet Russia, the substitution immigration into Palestine was furthered by close cooperation between representatives of the British government, the World Zionist Organization and its Executive in Palestine, the Federation of Jewish Labour in Palestine, and representatives of the Zionist movements still active in Soviet Russia. The present volume explores the motives and modes of operation of these many actors, dividing the discussion into two sections: one exploring the arrangements and mechanisms that enabled Zionist prisoners to leave Soviet Russia; the second surveying the often shaky cooperation that made possible their immigration into Palestine. To introduce the accounts of the substitution immigration, it is necessary to understand the history of Zionism during the early years of Soviet rule in Russia – its immense popularity among young Jews, surprising ability to survive and expand during the early years of Bolshevik dictatorship, and the organizational and ideological forms it adopted in these unusual circumstances.

The Zionist movements of the 1920s owed their vitality, in part, to events and developments in the years and decades preceding the establishment of Bolshevik rule and the crisis of Jewish life in Soviet Russia. In the early 1880s, prior Jewish hopes for gradual acceptance into a modernizing Russian society were brutally disappointed when the assassination of Tsar Alexander II led to a dramatic increase in anti-Jewish sentiment and the first wave of pogroms in two centuries. From that time, Russia became the site of mass Zionist and proto-Zionist sympathies, which served to catalyze the bulk of immigration into Palestine in the 1880s and in 1903–1914. Zionist influence became dominant among Russian Jews following the democratic revolution that swept away the tsarist regime in February 1917. The revolution relieved the Jews of the legal restrictions imposed on them by the tsarist regime, thus enabling them to participate in public life, to organize and publish. Yet, Russian Jewry continued to suffer from dislocation and poverty brought on by the mass expulsions of Jews from the western border provinces during the First World War. For Jews, the revolution's promise of universal democracy could not obliterate the message of national salvation, which took on a messianic fervour after the Balfour Declaration (November 9) and its promise of British support for a Jewish 'national home' in Palestine. At the end of 1917, membership in the Organization of Russian Zionists stood at 300,000, organized in 1,200 local branches.[2] The Zionist parties received more than two-thirds of the votes given to Jewish parties in the elections to the Russian Constituent Assembly.[3] Zionist associations, publications and cultural institutions appeared in every city, town and village, thus laying the organizational and emotive foundations for the persistence of Zionism over the coming decade.

The Bolshevik seizure of power in October 1917 in the name of a 'proletarian revolution' marked the beginning of a transition from the open political arena of 1917 to a dictatorship. The next few years were dominated by a bloody Civil War and foreign intervention, with disastrous consequences for every facet of life. The disruption of transportation and communications and the Bolshevik policies of nationalization, prohibition on commerce and requisitions caused endemic suffering and poverty. Moreover, the effects of the autarchic subsistence economy of the Civil War period were especially severe on the traditional Jewish occupations in trade and crafts. Some 70 or 80 per cent of Jews were without any source of livelihood, many engaging in speculation and black marketeering.[4] Worse yet, Jews were subject to the most horrific violence: more than two thousand pogroms took place in Ukraine and Belorussia between 1918 and 1920, most of them perpetrated by the anti-Bolshevik forces, leaving some 150,000 dead from direct violence and disease, another 500,000 homeless and 300,000 orphaned.[5] The pogroms put in doubt the physical survival of the Jews in Russia, whereas Bolshevik abolition of Jewish communal institutions (*kehillot*) and the attack on Hebrew language education and organized religion threatened their national and religious existence.[6] The situation paralyzed the Zionist Organization and its affiliates, but Jewish despair about survival in Russia and the search for salvation elsewhere only increased.

The end of the Civil War and the foreign intervention brought some relief to the Jewish communities in the western and southern reaches of Soviet Russia, but was slow to alleviate the burden of homelessness, hunger and unemployment. The New Economic Policy (NEP), established by the Soviet leadership in the early 1920s, removed many of the wartime restrictions and allowed private initiative in small scale production of consumer goods and in commerce. Many Jews used the new freedom of movement to leave their small towns and villages for the big cities of the former Pale of Settlement and the metropolises of southern and central Russia, finding work in the bureaucratic structures erected by the Soviet regime or entering institutions of higher education. But millions continued to live in the traditional small towns and were dependent on crafts and commerce, and these small Jewish operators were undermined by the high taxes imposed on their trades. Moreover, as 'exploiters' of the labour of others, they were labelled *lishentsy*, that is, deprived of the right to vote as well as several other crucial rights, including access to post-elementary education and to medical services and membership in the trade unions and their employment bureaus.

At the height of the relatively comfortable years of NEP in the mid-1920s, roughly one million Jews had no steady source of livelihood.[7] Policies launched in 1925 by the authorities to ease the economic plight of poor Jews through 'productivization' and 'agrarization' (i.e. through transition to work in crafts and industries as well as settlement on the land) had some salutary effect among the Jewish population; by the early 1930s, about a quarter-million Jews had moved into agricultural settlements in Crimea, Ukraine and Belorussia.[8] Yet, in the late 1920s, as NEP reached its end, approximately one-half to three-quarters of the Jewish youth in towns and cities with dense Jewish populations remained outside all existing educational and occupational frameworks.[9]

Zionism drew much of its appeal from the widespread sense of despair and marginality among Jews in the traditional Jewish regions, a feeling that was reinforced by the evolving policies of the Soviet government towards the nationalities making up the Union of Soviet Socialist Republics formally established in 1923. The high measure of national–cultural autonomy given to the 'territorial' nationalities within that union had practical and symbolic repercussions for the Jews. In Ukraine and Belorussia, the preference given to the national élites in governmental, educational and cultural institutions deprived some Jews of their earlier occupational and educational gains. More broadly, it highlighted the problematic status of the Jews as a 'non-territorial' nationality and undermined their confidence in a possible Communist solution to the persistent 'Jewish question'.

Two organizational forms predominated in Zionist activity during the 1920s: the youth movements and Hechalutz (Pioneer or Vanguard), an organization dedicated to training young Jews in physical labour, especially in agriculture, and in preparation for settlement in Palestine. The success of these organizations rested on their capacity to meet the needs of young Jews, to make adjustments to the norms of public activity in Soviet Russia, and to take advantage of the organizational opportunities opened up under NEP.[10]

The popularity of the youth movements resulted to a large extent from the failure of Communism to offer a satisfactory outlet for the needs and aspirations of many young Jews – those who could neither take advantage of new educational opportunities nor find employment, or were barred by their status of *lishentsy* from entering Communist youth organizations. True, the organizers of the youth movements came usually from among the growing numbers of Jewish students and many of them had been active in Jewish societies of university and gymnasium students during the World War and 1917. But the crisis of Jewish life in the years following October 1917 and the attack on established Jewish organizations propelled some among the educated youth into increased activism. Perceiving themselves as a new national leadership, their rhetoric and activity expressed deep empathy for the less fortunate among their fellow young Jews and the belief that true 'productivization' of the Jews could happen only through national rebirth in Palestine. To their young members (ages 17–23), the Zionist youth organizations offered opportunities for fellowship, ideological self-definition, and social commitment – as well as labour organization of working youth – that is, modes of being and acting that were promoted in the Soviet public arena of the time. Both the youth organizations and the children's movements they created (ages 10–16) emphasized educational activity including Jewish history, Hebrew culture and information about the Land of Israel, thus serving as something of a substitute for outlawed Hebrew language education and gaining the support of many Jewish parents. More than any other form of Zionist organization, the youth movements were protected from the Soviet system of political control since, by law, those under 18 could not be held in prison.

Hechalutz drew its inspiration from notions developed during the decade before the First World War by young Russian immigrants who came to Palestine as pioneers of Jewish labour and the would-be builders of a new Jewish society. Their experience taught them the need for thorough training (*hakhshara*) of future immigrants in physical labour and agriculture. This was the aim of the farms and workshops established by Hechalutz from its inception in 1918. But at that time and in coming years, the training enterprises also answered a practical need of many young Jews who lacked a livelihood or a roof over their heads. Membership in these communal farms and workshops (whose numbers grew fast during the years of NEP) also promised a sense of belonging and the hope of collective national betterment. Like the youth movements, these communal enterprises were in step with the current policies and initiatives of the Soviet government – a coincidence that helped them survive and diminished the inner conflict experienced by those of their members who saw themselves as socialist or were otherwise sympathetic to the Soviet project. An additional measure of protection from Soviet censure as well as direct financial aid came from the US-based Joint Distribution Committee (JDC), which supported all forms of Jewish agricultural settlement in Soviet Russia.[11]

Politically and ideologically, all of the Zionist organizations of the 1920s issued from the populist, 'democratic' wing of the Russian Zionist movement. Difficult living conditions, Soviet intolerance towards alternative political organizations and the attack on religion, all contributed to the decline of both religious Zionism and the 'General Zionists' who had dominated the movement in 1917. By 1920, the democratic wing itself, known as the 'Popular Faction – Tze'irei Tziyon', split into two parties: the Zionist–Socialist Party (TzS – an acronym from the Yiddish) and the Zionist Toilers' Party (STP in the Russian acronym, Tze'irei Tziyon-Hitachdut in Hebrew).[12] The parties disagreed bitterly over their respective attitudes towards the Soviet revolution and the role of socialism in the national revival of the Jews, though neither party developed a clear, inclusive doctrine. TzS viewed Zionism and socialism as inseparable. Theirs was an eclectic brand of socialism, coloured by their admiration for the voluntarism of the Bolsheviks. They rejected the Marxist doctrine held by the Bolsheviks and condemned Bolshevik dictatorship, yet accepted the 'Soviet platform', that is, the right of the workers and their councils (or soviets) to establish a dictatorship in order to fight the enemies of the proletarian revolution. Much of the activity of TzS was conducted among the impoverished and unemployed in behalf of their 'productivization' and organization in cooperatives and trade unions. STP objected to the symmetry drawn by TzS between socialism and Zionism and rejected every aspect of the Soviet regime and the social organizations it created. In their discussions of the national Jewish renaissance in Palestine, the leaders of STP proclaimed the primacy of the national task over the social, although they, too, gave central place to the transformation of the Jews into labourers.

The split between socialists and toilers affected every Zionist movement and organization operating in Soviet Russia. Both parties attempted to win over the Jewish youth and the membership of Hechalutz. They were helped from inside the movements by the increasing disagreements between socialists of various shades and their opponents. By 1924, there emerged some six or seven youth and children's movements. STP worked to unite the national wings of several student movements in one organization known as EVOSM (the Russian acronym for the United All-Russian Organization of Zionist Youth). Members of EVOSM began, in 1924, to organize younger cohorts into the children's movement Ha-Shomer ha-Tza'ir ha-Amlani (the Toilers' Wing of The Young Guard). TzS organized its own youth movement, TzS Yugend Farband (the Zionist Socialist Youth League).

which was active primarily among working youth. Also operating within the ideological orbit of TzS was Ha-Shomer ha-Tza'ir ha-Ma'amadi (the class wing of The Young Guard), a movement built by Jewish students on the model of scouting movements, embracing both children and youth. Two other youth movements of strong socialist convictions, Dror and ESSM (Russian acronym for the Jewish Socialist Youth League) had strong followings in Moscow, Leningrad, Kiev and Minsk.

In Hechalutz there also developed two ideological and political trends, paralleling the eclectic socialism of TzS and the 'toiler' position of STP. Internal divisions sharpened during the final stages of a long campaign for legal recognition led by the socialist majority of the organization's Central Committee. The formal split (autumn 1923) came within weeks of the registration of Hechalutz as a recognized public organization, with its headquarters in Moscow. Hechalutz was also allowed to publish a journal, though the publication began to confront difficulties after its first year and was discontinued in 1926.[13] Those opposed to the registration formed their own organization – 'Illegal' or 'National' Hechalutz. They rejected any adaptation to Soviet conditions, which they viewed as hostile to the Jewish national rebirth and capable of tempting and misguiding Jewish youth. The 'Legal' or 'Class' wing continued to advocate activity within the confines of Soviet dictatorship and argued that the Zionists shared with the Soviet regime the goals of Jewish productivization and a progressive Middle East. Some of the sharpest debates between the two wings concerned the form of settlement that should prevail in Palestine – the left wing insisting on the 'collectivist' principle (i.e. the establishment of *kibbutzim* or communal agricultural settlements) while the right wing supported loose cooperative settlements (*moshavei ovdim*).

Notwithstanding the ideological and organizational divisions, the years 1923 through late 1925 (and in some cases mid-1926) saw Zionist influence expand rapidly in areas of dense Jewish population and in big cities throughout the former Pale of Settlement, the South and in Central Russia. By 1925, some ten Zionist parties and organizations, each with a national network of branches and cells, claimed a combined membership of several tens of thousands.[14] These numbers, and the steady growth of the movements, testify not only to the attraction of Zionism for many young Jews, but also to a surprising organizational latitude enjoyed by the Zionists during those years.

To be sure, the Zionists operated within a self-declared political dictatorship. Moreover, Zionism had its detractors, foremost among them was the Jewish Section of the Communist Party (Evsektsiia). Soon after its birth in the early days of the Civil War, the Evsektsiia began to press the Party and the political police (known at different times as the Cheka, GPU and OGPU) to ban all forms of Zionist activity. As the Zionist movements expanded, the Evsektsiia complained bitterly that its Communist operatives were at a grave disadvantage in the Jewish arena.[15] At times, the Evsektsiia's pressure brought results: in July 1919, a short-lived Bolshevik regime in Ukraine proclaimed the dissolution of all Zionist organizations; in September 1919, the Cheka arrested the Central Committee of the Zionist Organization in Petrograd; in April 1920, all delegates to the Zionist

Conference in Moscow were arrested; in August 1922 branches of Hechalutz suffered searches and arrests: in March 1924, the GPU in Moscow rounded up leaders of several Zionist organizations. Each of these incidents was preceded by strenuous Evsektsiia lobbying, as were many of the local incidents of harassment and arrest.[16] Yet, until 1923, such arrests ended in quick release and did not as a rule lead to imprisonment or exile. Only a handful of people were exiled to remote places in 1923 – mostly leaders of the Zionist Socialist Party – arrested in Ukraine after the republic's authorities rejected their request for legal recognition.[17] At other times, Zionists were caught up in broader sweeps of political opponents, as happened in May 1922, when the delegates to the founding conference of STP in Kiev were arrested and put on public trial – much like the Socialist Revolutionaries and Orthodox Church leaders at that time.[18] Unlike the latter, their sentences of internal exile were commuted in autumn 1923 to 'deportation' from Soviet Russia.

However, the most remarkable factor permitting the rapid increase in Zionist activity was the relative tolerance shown by the Soviet authorities towards the Zionist organizations, especially the youth movements and Hechalutz. Political leaders were often arrested, and local authorities occasionally harassed and detained groups of Zionists in their area, but there was no concerted, unified effort to demolish the organizational base of the Zionist youth organizations or even the parties.[19] Hechalutz farms, whether 'legal' or 'illegal', continued to operate largely unhindered until 1926. Two formal acts reflected this tolerance and provided the Zionists with a measure of legal protection. A decision of the All-Russian Executive Committee of Soviets (VTsIK) on 21 July 1919 instructed all Soviet organizations not to place obstacles in the path of the Zionist Organization of Russia – 'to the extent that [its] cultural and educational activities do not contradict the resolutions of the Soviet State' – since it had not been identified by either VTsIK itself or the Council of People's Commissars (Sovnarkom) as counterrevolutionary.[20] More conclusive and effective was the registration of Hechalutz by the People's Commissariat for Internal Affairs (NKVD) – essentially legalizing it – under instructions from Feliks Dzerzhinskii as NKVD head and Stalin in his role as Commissar for Nationalities.[21]

Conditions for Zionist activity began to worsen in 1924 and drastically so in the middle of 1925. There was more than one reason for this change. A significant factor, initially, was the process of radicalization that unfolded within the Zionist youth movements. Throughout their years of expansion, these movements competed with the Evsektsiia, Committee for Agricultural Settlement of Toiling Jews (Komzet) and the Communist Youth League (Komsomol) for influence over the Jewish youth. They used the public forums made possible under NEP to sound their Zionist message, speaking at assemblies for 'unaffiliated youth', meetings of trade and youth organizations and rallies to mobilize support for the agricultural settlement project. Harassed by the Communists and occasionally arrested, the Zionist youth began to shift the focus of their rhetoric from Zionism as such to slogans directed against the Evsektsiia. A dramatic escalation in the war of words occurred on 16 August 1924, when the Zionist Socialist Party and its youth movement distributed 10,000 copies of a fiercely anti-Evsektsiia flyer.[22] The authorities responded with the first truly massive arrests on 2 September, when 3,000 Zionists were rounded up by the Ukrainian GPU. These arrests led to protests and then another wave of arrests in early 1925, although even then, less than 10 per cent of those arrested received sentences of exile.[23] By the summer of 1925, all the youth movements, regardless of their earlier position on the 'Soviet platform', resorted to decidedly anti-Soviet slogans, thus exposing themselves to the full force of Soviet dictatorship.[24]

Meanwhile, at the highest reaches of the Soviet leadership, power began to shift from one alliance to another. While Stalin continued to accumulate power, several Bolshevik leaders in whom the Zionists had found a sympathetic ear were no longer in the innermost sancta of decision making. Among them were Lev Kamenev, Aleksei Rykov, Petr Smidovich, Iurii Larin, Nikolai Krestinskii and Ruben Katanian.[25] Feliks Dzerzhinskii, the head of the OGPU and a frequent advocate of tolerance towards the Zionists,[26] directed more of his attention to his duties as head of the All-Union Council on National Economy. By mid-1925 he had been struck by the illness that would bring on his premature death, thus making way for subordinates who had called for tougher punitive measures from the outset. The decline of these leaders coincided with increased pressure against the Zionists from the authorities in Ukraine and Belorussia, who feared the growing influence of the Zionists in their 'national' republics as is attested, for instance, in Eduard Karlson's memorandum cited in the opening of this chapter. Pressure also came from the organization charged with the agricultural settlement of the Jews, Komzet. Always fearful that unfavourable sentiment in the Jewish population could disrupt the Soviet 'agrarianization' campaign, it now came to view the repression of the Zionists as less damaging than their continued activity. By 1927, when the Soviet leadership raised the idea of an autonomous Jewish region in Birobidzhan, the Zionist message of emigration to Palestine became an intolerable affront to Soviet policies and claims.

The shift in Soviet treatment of the Zionists is documented in the diary of the man who had served all along as the unofficial intermediary between the Zionists and the more sympathetic Soviet leaders. This was Professor David Shor, a renowned pianist in pre-revolutionary Russia and, from 1918, a leading teacher and mentor at the Moscow Conservatory of Music. His personal friendship with Kamenev and those in his political circle was instrumental in allowing various Zionist leaders' access to the authorities (including a crucial meeting between representatives of Hechalutz and Stalin, which paved the way to the 'legalization' of that organization). For added influence, he recruited the help of Dr Joseph Rosen, who directed the JDC's affairs in Moscow. In spring 1925, Shor held promising discussions with Kamenev, Rykov, Smidovich and Larin concerning the establishment of an emigration bureau in Moscow to regularize emigration to Palestine. These discussions – and the relatively tolerant treatment of Zionists – came to an end in a fateful meeting between Shor, Smidovich and two GPU officials, where the latter presented numerous examples of anti-Soviet flyers issued by the Zionist youth movements.[27]

Organized Zionism did not die immediately. Hechalutz retained its legal status until 1928, when representatives of the Evsektsiia and Komzet served its leaders with a closing order and took over the organization's office in Moscow. By then, however, the two wings of Hechalutz had lost many of their members. Whether 'legal' or not, their farms were an easy target for the security services and agencies dealing with agricultural settlement. Starting in 1926, groups of members were driven out of Tel Chai, Mishmar and other Hechalutz farms in Crimea and elsewhere. In 1927, Komzet deprived all Jewish collective farms of the right to choose their members, thus effectively taking control of the farms' management. Events in Palestine added to the demoralization of Hechalutz members, as economic crisis and unemployment brought about suspension of immigration. Only financial help from abroad and the return of two leaders of Illegal Hechalutz from Palestine (both eventually arrested) allowed remnants of that organization to survive until 1929.[28]

Of the youth movements, the large organizations working among older youth – EVOSM and TzS Yugend – had already disintegrated in the first half of 1926, weakened by arrests and the demands of underground work. But the two wings of Ha-Shomer ha-Tza'ir (encompassing children as well as youth) survived considerably longer, supported by reserves of younger members and the ability of the leading cadres to conduct work underground. The 'national' wing transformed itself in 1927 into small, 'closed' communes, loosely united into 'labour brigades' in the big cities of central Russia, where they were relatively inconspicuous. As late as 1934, the Moscow 'brigade' reported 100 members.[29] The 'class' wing of Ha-Shomer ha-Tza'ir continued to operate on a large-scale longer than any other movement, reporting as many as 6,000 members in the middle of 1927, coordinated through 13 regional committees. Their survival was made possible by an energetic and effective command, composed largely of leaders who had escaped from exile and now operated underground. The last known communication from the movement's 'General Staff' was sent from Moscow in January 1933,[30] though a group of members who had completed their terms of internal exile and settled in Kalinin, attempted in 1934 to revive contacts with members elsewhere. All were soon arrested.[31]

By the mid-1930s, organized Zionism had ceased to exist in the Soviet Union. But the unique organizational and ideological patterns that had sustained it during the 1920s did not entirely disappear. At least some were transplanted into Palestine by nearly three thousand committed Zionist immigrants who arrived from Soviet Russia between 1924 and 1934, first and foremost among them, the many hundreds arrested for Zionist activity and allowed to substitute their terms of exile with permanent departure for Palestine.

The substitution emigration was not without precedent. In January 1922, several imprisoned Menshevik leaders were allowed to leave Soviet Russia in lieu of internal exile; later that year, many intellectuals and academics arrested for publicly voicing criticism of the regime were forcibly deported abroad; in autumn 1923, the leaders of Tze'irei Tziyon who had been arrested in 1922 and sentenced to exile were allowed to leave for Europe; later on, in 1926, the old Menshevik leader A.N. Potresov was permitted to go abroad to seek medical help, on the tacit understanding that he would not return. Still, the substitution arrangement under which the Zionists left was unique in origin, duration and effect. It was initiated by Professor David Shor. In spring 1924, following the arrest of leaders of several Zionist movements, twelve of whom were ordered exiled to the Narym region in Siberia, he called Ruben Katanian, the General Prosecutor, and pleaded to have them sent instead to Palestine. He also spoke to Kamenev and Smidovich and arranged for the prisoners to be released from prison while awaiting their departure for Palestine in July 1924.[32] Like the Menshevik leaders in early 1922, the Zionists were allowed to choose between exile and departure. Whereas the famous deportations of the intellectuals had been forced upon them, the Zionists' exit was voluntary.

Shor's *ad hoc* initiative laid the foundation for a complex set of arrangements which lasted an entire decade. The Soviet motives in allowing the Zionists to leave for Palestine changed over time. Initially, the goal was to remove the most active Zionists from the scene without sending them to prison or exile, that is, without upsetting the Jewish population and the JDC. Even after the Zionist movement had been radically reduced, emigration to Palestine continued to be used as a safety valve, this time to blunt criticism from international Jewish organizations and foreign governments.[33] Moreover, a dramatic increase in the price of the required exit documents turned the emigrants into a source of revenue for the Soviet government.[34] Still, it is remarkable that the substitution arrangements remained in force until 1934, that is, through Stalin's rise to near absolute power and the implementation, under his leadership, of ruthless policies of collectivization and industrialization. As will be detailed in the section on 'Substitution' in Chapter 2 of this volume, the Soviet authorities put many obstacles in the way of prospective substitution emigrants, and many failed to realize this option. Nevertheless, some 1,200 or 1,300 – roughly one half of those sentenced by the Soviet authorities – were able to escape political repression and leave for Palestine.[35] Their success in doing so depended first and foremost on the help of Ekaterina Peshkova and her Committee to Aid Political Prisoners. In addition, the consummation of this unusual substitution required cooperation and support from several actors outside the Soviet Union; their role will be examined in the section on Pompolit in Chapter 2 of the volume. Finally, the effects of the substitution emigration on those who benefited from it and on the society they joined in Palestine will be explored in a postscript and contrasted with the many Zionists whose fate – or choice – was to spend their lives in Soviet Russia.

Notes

1 Translated and introduced by Mattityahu Mintz, 'Illegal Zionist Organizations in Ukraine, 1924–1925', *Jews in Eastern Europe*, 3 (Winter 1996), pp. 46–66.
2 Ar'ye Refa'eli (Tzentziper), *Ba-Ma'avak li-Ge'ula. Sefer ha-Tziyonut ha-Rusit mi-Mahapekhat 1917 ad Yameinu* (Tel Aviv: Davar and Ayanot, 1957), p. 29. Yitzchak Ma'or, *Ha-Tnu'a ha-Tziyonit be-Rusya me-Reshita ve-ad Yameinu* (Jerusalem: Ha-Sifriya ha-Tziyonit and Magnes, 1973), pp. 460–89.

3 Oliver H. Radkey, *The Elections to the Russian Constituent Assembly of 1917* (Cambridge, MA: Harvard University Press, 1950), p. 17; Zvi Y. Gitelman, *Jewish Nationality and Soviet Politics. The Jewish Section of the CPSU, 1917–1930* (Princeton, NJ: Princeton University Press, 1970), pp. 79–80.
4 Ya'akov Leshchinski, *Ha-Yehudim be-Rusya ha-Sovyetit mi-Mahapekhat Oktober ad Milkhemet ha-Olam ha-Shniya* (Tel Aviv: Am Oved, 1943), p. 69.
5 Gitelman, *Jewish Nationality and Soviet Politics*, pp. 159–68; Leshchinski, *Ha-Yehudim be-Rusya ha-Sovyetit mi-Mahapekhat Oktober*, p. 53.
6 A decree on the dissolution of the *kehillot* was published in summer 1919, although, as pointed out by Gitelman, many *kehillot* continued to operate under various guises well into the 1920s. Gitelman, *Jewish Nationality and Soviet Politics*, pp. 271–2.
7 This figure is cited by Leshchinski, *Ha-Yehudim be-Rusya ha-Sovyetit mi-Mahapekhat Oktober*, pp. 86–8, and is based on the 1926 census.
8 Leshchinski, *Ha-Yehudim be-Rusya ha-Sovyetit mi-Mahapekhat Oktober*, p. 164.
9 Ibid., pp. 100–3, cites several official publications and articles in the Soviet press.
10 The following published writings by veterans of the Zionist movements were drawn on for the discussion of these organizations: Ar'ye Refa'eli (Tzentziper), *Ba-Ma'avak li-Ge'ula, 1957*, and *Eser Shnot Redifot. Megilat ha-Gzeirot shel ha-Tnu'a ha-Tziyonit be-Rusya ha-Sovyetit* (Tel Aviv: Ha-Ve'ada ha-Historit shel Brit Kibbutz Galuyot, 1930); Yitzchak Rabinovich, *Mi-Moskva li-Yerushalayim* (Jerusalem: Re'uven Mas, 1957); Yehuda Erez (ed.), *Sefer TzS. Le-korot ha-Miflaga ha-Tziyonit-Sotzialistit u-Brit No'ar TzS be-Brit ha-Mo'atzot* (Tel Aviv: Am Oved, 1963); Israel Ritov, *Prakim be-Toldot 'Tze'irei Tziyon'-TzS* (Tel Aviv: Am Oved, 1964); Zalman Aran, *Otobiografiya* (Tel Aviv: Am Oved, 1971); Binyamin Vest (ed.), *Naftulei Dor. Korot Tnu'at ha-Avoda ha-Tziyonit Tze'irei Tziyon-Hitachdut be-Rusya ha-Sovyetit* (Tel Aviv: Mishlachat Chul shel Tze'iri Tziyon–Hitachdut be-Rusya, 1945); *Hechalutz be-Rusya. Le-toldot Hechalutz ha-Bilti-Legali (ha-Le'umi-Amlani)* (Tel Aviv: Mishlachat Chul shel Hechalutz ha-Bilti Legali be-Rusya, 1932); Moshe Basok (ed.), *Sefer Hechalutz, Antologiya* (Tel Aviv: Ha-Sokhnut ha-Yehudit-Machleket ha-Aliya, 1940); Yehuda Erez (ed.), *Chalutzim Hayinu be-Rusya* (Tel Aviv: Am Oved, 1976); Dan Pines, *Hechalutz be-Kur ha-Mahapekha. Korot Histadrut Hechalutz be-Rusya* (Tel Aviv: Davar, 1938); Lyova Levite, *Bereshit va-Sa'ar. Edut Ishit* (Tel Aviv: Ha-Kibbutz ha-Me'uchad, 1978), and *Machteret tziyonit-sotzialistit-chalutzit be-Brit ha-Mo'atzot, Shorashim, Kvatzim le-Cheker ha-Kibbutz u-Tnu'at ha-Avoda*, Vol. 5 (Tel Aviv: Yad Tabenkin and Ha-Kibbutz ha-Me'uchad, 1986), pp. 45–79; Avraham Tarshish (ed.), *Kovetz Dror. Min ha-Mahapekha ha-Rusit el ha-Mahapekha ha-Ivrit* (Tel Aviv: Yad Tabenkin and Ha-Kibbutz ha-Me'uchad, 1981); Avraham Itai, *Korot Ha-Shomer ha-Tza'ir be-S.S.S.R. No'ar Tzofi Chalutzi – NeTzaCh* (Jerusalem: Ha-Aguda le-Cheker Tfutzot Israel, 1981). Also used were the following archival collections: Hechalutz Archive: Gosudarstvennyi Arkhiv Rossiiskoi Federatsii (GARF), Moscow, fond f. 7747; materials of TzS and EVOSM at the Tzentziper Collection: Central Zionist Archives (CZA), Jerusalem, F30; materials of Hitachdut (STP) and Illegal Hechalutz at the Binyamin Vest Collection: Lavon Institute for Labour Research, Tel Aviv, IV-104-53.
11 In September 1923, David Ben-Gurion wrote from Moscow that nearly all of the budget of Hechalutz farms (about $10,000) came from the US-based JDC. *Igrot David Ben-Gurion*, Vol. 2 (1920–1928). Compiled and edited by Yehuda Erez (Tel Aviv: Am Oved and Tel Aviv University, 1973), p. 166.
12 The party's full Russian name was Sionistskaia Trudovaia Partiia. The split between socialist and toiler Zionists reflects a parallel division – both ideological and linguistic – between *sotsialisty* or *rabochie* and *trudoviki* in the Russian labour parties.
13 A large issue of *Hechalutz* (nos. 15, 16, 17) appeared in 1926, but most of the information included was from 1925. *Davar*, 13 September 1926.
14 In autumn 1923, Ben-Gurion reported that TzS and STP, each believed the number of its followers to be around 3,000: *Igrot Ben-Gurion*, Vol. 2, p. 168. The figures reported

by the largest youth movements and the two wings of Hechalutz in their retrospective documentation add up to roughly 50,000: 12,000 for EVOSM, 6,000 for Yugend, and 12,000 for the left wing of Ha-Shomer ha-Tza'ir, all in 1925. In 1926 – 9,000 in Illegal Hechalutz and 7,000 in the Legal organization. Vest, *Naftulei Dor*, pp. 75–6, 180; Refa'eli, *Ba-Ma'avak Li-Ge'ula*, p. 137; Itai, *Korot Ha-Shomer ha-Tza'ir be-S.S.S.R.*, p. 86; Pines, *Hechalutz be-Kur ha-Mahapekha*, pp. 199, 234; *Hechalutz be-Rusya*, pp. 36, 40, 42; Erez, *Chalutzim Hayinu be-Rusya*, pp. 491–4. In evaluating these figures, it should be remembered that there was considerable overlap between EVOSM and Illegal Hechalutz, and between TzS Yugend and Hechalutz's Legal wing. Available contemporary figures are somewhat lower. The newsletter of the world organization of Hitachdut reported in January 1925 that even after the large-scale arrests of autumn 1924 there were 2,000 members in the party, 2,000 in EVOSM, and 3,000 in Illegal Hechalutz. The formation of a non-socialist organization of Ha-Shomer ha-Tza'ir was noted but no membership figures were supplied: Lavon Archive, IV-104-53, file 55. At roughly the same time, reports at the Youth Conference and Hechalutz Council in Danzig, Germany enumerated the following membership numbers: TzS Yugend – 3,000; the 'class' organization of Ha-Shomer ha-Tza'ir – 8,000; EVOSM – 4,000: *Kuntres*, Vol. 9, No. 189, 9 October 1924, p. 16.

15 See, for example, Rossiiskii Gosudarstvennyi Arkhiv Sotsial'noi i Politicheskoi Istorii, Moscow (RGASPI), f. 445, op. 1, d. 40, ll. 157–8 and elsewhere in the Evsektsiia's archive (f. 445) at RGASPI.

16 RGASPI, f. 445, op. 1, d. 22, ll. 21–3, 33–40ob., 45–48; d. 119, ll. 16, 120–1, 146–7ob.; d. 167, l. 1; d. 179, l. 22; d. 192, ll. 2, 4–6, 7.

17 *Kuntres*, Vol. 11, No. 215, 8 May 1925, p. 31.

18 The arrest of the Zionist Central Committee in September 1919 also coincided with a renewed campaign of repression against opposition parties (especially the Mensheviks and Socialist-Revolutionaries) during an offensive by the anti-Bolshevik forces, whereas the September 1922 arrests came amidst a campaign against dissenting intellectuals and academics.

19 Other historians support this conclusion, at least partially. Schechtman writes that there was no 'massive and consistent anti-Zionist drive' in the early years, although he puts the beginning of such a drive in 1922. J.B. Schechtman 'The USSR, Zionism, and Israel' in Lionel Kochan (ed.), *The Jews in Soviet Russia since 1917* (Oxford: Oxford University Press, 1972), p. 105. G.V. Kostyrchenko in a recent, thoroughly researched study, presents a more ambiguous picture. He argues that the Communist leadership decided already in 1920 to destroy Zionism, yet did not move to do so in earnest until 1924, due to 'bureaucratic inertia' and possibly an intervention by the JDC. *Tainaia politika Stalina. Vlast' i antisemitizm* (Moscow: Mezhdunarodnye otnosheniia, 2001), pp. 70–1, 79. My own work in the archives of the Cheka/GPU and the Political and Organizational Bureaus of the Central Committee of the Communist Party show these bodies to have been divided among themselves over the danger posed by Zionism.

20 RGASPI, f. 17, op. 84, d. 44, l. 5.

21 About Stalin's involvement, see Pines, *Hechalutz be-Kur ha-Mahapekha*, pp. 189–92. For a copy of the official announcement of Hechalutz status see GARF, f. 7747, op. 1, d. 8, ll. 75–75ob.

22 Erez, *Sefer TzS*, pp. 102–4, 134–7.

23 TzS reported that 3,617 Zionists had been seized by the GPU between 20 March 1924 and 1 May 1925, but as of mid-1925, only 157 had been sentenced to exile and 160 were awaiting their sentence. *Davar*, 3 July 1925. These figures are within the same range as those mentioned in a GPU report from 29 May 1925: 13 given three-year terms in *kontslagery*; 132 sent into exile, most of them for three years; 34 in prisons, presumably awaiting sentencing; 152 had received substitution or had been sent abroad (the later category presumably referred to the Tze'irei Tziyon delegates arrested in 1922 and sent abroad in early 1924). RGASPI, f. 76, op. 3, d. 326, l. 5.

24 See, for example, the broadsheets produced by TzS Yugend, EVOSM, and the class wing of Ha-Shomer ha-Tza'ir in Kiev, Kremenchug, Minsk and other locations: RGASPI, f. 445, d. 167, ll. 51–56; f. 17, op. 84, d. 748, ll. 129–33.
25 Lev Kamenev was arguably the strongest Bolshevik leader at that time, taking Lenin's place as Chairman of the Sovnarkom and at the head of the Politburo. Aleksei Rykov served as Deputy Chairman of Sovnarkom; Petr Smidovich was the Deputy Chairman of VTsIK and the head of Komzet; Iurii Larin's duties included chairmanship of Komzet's public arm, the All-Union Society for Agricultural Settlement of Toiling Jews in the USSR, better known under the acronym Ozet; Nikolai Krestinskii headed the Commissariat for Finances; Ruben Katanian was the General Prosecutor.
26 For an English translation of Dzerzhinskii's letters to his underlings on this theme, see M. Beizer and Vladlen Izmozik, 'Dzerzhinskii's Attitude toward Zionism', *Jews in Eastern Europe*, 25 (Spring 1994), pp. 64–70 and Documents 4 and 7 in this volume, which have been retranslated here from the Russian.
27 Shor wrote about this in his diary entries for 11 March 1925 and 16 June 1925: The Shor Family Collection, Manuscript Collections of the National and University Library (NUL), Jerusalem, 1521/449/6, p. 250; 1521/450, p. 252. His information is corroborated in two contemporary press reports: *Davar*, 19 July and 10 August 1925.
28 *Hechalutz be-Rusya*, pp. 43–6; Pines, *Hechalutz be-Kur ha-Mahapekha*, p. 244; Erez, *Chalutzim Hayinu be-Rusya*, pp. 458–62, 464–5. While in operation, Legal Hechalutz's Centre helped the immigration to Palestine of many of its members who were driven out of Tel Chai and other farms. Groups of immigrants are reported to have left in late 1924, in winter 1925–1926, and in 1928 and 1929.
29 Vest, *Naftulei Dor*, pp. 71–3, 129–32.
30 Itai, *Korot Ha-Shomer ha-Tza'ir be-S.S.S.R.*, pp. 258–68.
31 Introduced by Michael Baizer and Arkadii Zeltser, 'Hashomer HaTza'ir in Kalinin, 1934', *Jews in Eastern Europe*, 3 (Winter 1997), pp. 51–71. The movement's own history puts the last meeting held by its members (including some from Kalinin) in March 1935 in Tobol'sk. Itai, *Korot Ha-Shomer ha-Tza'ir be-S.S.S.R.*, pp. 279–81.
32 Shor, NUL 1521/449/5, pp. 91–2, 94–5.
33 As discussed in Chapter 3 of this volume, the Zionists used contacts with the British and US governments.
34 See more on this issue in Chapter 2: Out of the Soviet Union.
35 For a discussion of the information used to establish the number of substitution immigrants (as well as the number of those who did not benefit from substitution) see the final pages of Chapter 3: Into Palestine, in this volume.

2 Out of the Soviet Union
The exiles and Pompolit

Boris Morozov

Arrests and internal exile

The Bolshevik leadership was opposed in principle to the idea of creating a 'Jewish national enclave' in Palestine. As early as 1903 Lenin declared that, '...this Zionist idea is entirely false and reactionary in its essence'.[1] After the 1917 revolution, however, the state tolerated Zionist activity. Thus, in 1919 the All-Russian Executive Committee of Soviets (VTsIK) declared that

> ...since VTsIK and Sovnarkom have not in a single decree declared the Zionist Party to be counterrevolutionary, to the extent that [its] cultural and educational activities do not contradict the resolutions of the Soviet State, the presidium of VTsIK instructs all Soviet organizations to refrain from creating any obstacles to this party in [its] activities...[2]

Arrests and persecution occurred in the early period following the revolution, but initially repressive measures against Zionists were of a sporadic and unsystematic nature. Authorities in Vitebsk closed a cell of Hechalutz in July 1919;[3] in January 1920 Zionists were arrested in Zhitomir and in April arrests were made in Odessa for publication and distribution of Zionist leaflets. Later that year Zionist activists were detained in Samara and Ekaterinburg.[4] Members of Hechalutz were arrested and accused of aiding illegal emigration to Palestine in 1921 in Odessa,[5] while in Minsk the authorities decided to crack down on the Zionist student organization, Hechaver.[6]

Even when arrests were made, the sentencing reflected the ambiguity of the official stance on the Zionist question. On 23 April 1920, in Moscow, all delegates present at the All-Russian Zionist Conference were arrested, sent to the Lubianka prison and indicted for holding the conference without proper permission. Those younger than 20 and older than 60 – ten altogether – were released after a brief interrogation. The remaining detainees were imprisoned first in the Lubianka internal prison and then transferred to Butyrskaia prison until 16 May. Between 16 May and 21 June they were gradually released with the exception of nineteen people, for whom a sentence was handed down on 29 June. Seven of these were accused of counterrevolution and opposition to the regime and received five years forced labour. *Izvestiia* made the dubious claim that the accused had colluded with the allied intervention into Soviet Russia in 1919–1920 in exchange for Jewish rights in Palestine. After the sentence went into effect on 2 July, all nineteen were freed.[7]

In 1922–1923 the picture began to change: the number of arrests of members of Zionists organizations increased and harsher sentences were implemented. Members of the Hechalutz movement were arrested in Kharkov, Gomel, Kiev and Odessa. Arrests of members of other Zionist organizations such as the Zionist Toilers' Party, Maccabi and the Zionist Socialist Party (TzS) were carried out in the Crimea,[8] Gomel Province, Moscow, Petrograd, Kiev, Novonikolaevsk, Irkutsk and other cities.[9] While many of those arrested during this period were released after several days, relatively severe punishments were also meted out. For example, delegates to the April 1922 conference of Tze'irei Tziyon were arrested and put on public trial in August 1922 – the only one of its kind aimed at Zionists – and about a third of them served sentences in prison and internal exile before being deported.[10]

The crackdown against the Zionists in this period seems to have stemmed from an overall upsurge in their activity. According to a classified report of the Secret Department of the State Political Directorate (GPU)[11] on searches of activists in various Zionist organizations in 1923, overwhelming evidence was found of 'intensified activities on the part of these anti-Soviet illegal organizations'.[12]

A turning point came in 1924. Thousands of Zionists were seized in a series of arrests which took place in Moscow, Leningrad and practically all the large cities of Belorussia and Ukraine. This wave culminated on 2 September when, in the course of a single day, more than 3,000 Zionists were arrested in 150 locations.[13]

From 1925 to 1929 practically all remaining activists and rank and file members of all Zionist organizations were detained. In the early 1930s, many of the Zionists who had been exiled were rearrested by the United State Political Directorate (OGPU) immediately upon completion of their sentences, and by the middle of that decade, the Zionist movement in Russia had, to all intents and purposes, been eliminated.

This is particularly striking considering that through 1925, OGPU Chief Feliks Dzerzhinskii still held firmly to the view that the disadvantages of persecuting the Jews outweighed the advantages. 'I have reviewed the Zionist materials', he informed his subordinates, Viacheslav Menzhinskii and Genrikh Iagoda on 15 March 1924, and 'I must admit, I do not entirely understand why they are being persecuted for their Zionist affiliations... Persecuted, they are a thousand times more dangerous to us than were they not persecuted...'[14]

Operational responsibility for control of Zionist organizations fell within the jurisdiction of the 10th (after January 1925, the 4th) section of the Secret Department of the OGPU, which was headed in 1922–1923 by T.P. Samsonov, and after 25 May 1923 by Terentii Dmitrievich Deribas. Deputy Director Yakov M. Genkin was responsible for day by day operations.[15]

The OGPU's *modus operandi* was to converge on groups of Zionists who had gathered for congresses, conferences or other organized activities.[16] It was typically the most active members of the organization, who were arrested, although there were also cases in which, for example, an entire room full of delegates to a convention of a Zionist movement were detained. Given the ambiguities regarding the legality of their activity, the safest approach was to charge them with failing to receive permission to organize these events. Charges of 'anti-Soviet activity' were often tacked on afterwards. Those arrested were usually detained in a OGPU prison where they underwent further interrogation. In the early years of 1922–1923, Zionist detainees were often held together, which enabled them to coordinate their testimony and agree on strategies of behaviour under interrogation. This probably reflected haphazard organization in the fledgling security services as much as it revealed perceptions regarding the level of threat that Zionism posed to the state.

There were also cases in which individual activists were singled out and pursued by the OGPU; their activities were monitored, their premises were searched and they were ultimately arrested.[17]

Initially, arrests were made only after the discovery of evidence proving active involvement in Zionist organizations. Later the OGPU dispensed with the formality of finding just cause for detention, and the blatant order given was, 'search and arrest'.[18]

Detainees were subjected to lengthy and exhausting nocturnal interrogations which, at times, involved provocations and psychological pressure.[19] However, in the early 1920s, unlike the period of the Great Terror, the behaviour of investigators was relatively 'proper'. Physical torture was rare, and conditions of detainment were tolerable.[20]

The initial task of the investigators in the majority of cases was to establish the identity of those arrested. Many Zionists who had worked illegally were arrested with false documents in hand or without any documents at all. Some were tried and even exiled under assumed names. The next stage was to extract an admission of participation in a Zionist organization. Pressure was then exerted on the detainee to renounce this membership. Although there does not seem to have been a standard form for such a declaration, the texts were similar. The signatory acknowledged the anti-Soviet nature of Zionist activity and renounced his current and any future involvement in Zionist organizations. A person who signed such a declaration was usually released. Naturally, the signature was perceived by other Zionists as a betrayal. In order to coax activists to sign, the OGPU offered empty assurances that the declaration would be kept secret. However, in certain periods regulations actually required that these statements be made public, and even when not required, the declarations were usually passed along to the periodical press as a matter of course, disgracing the signatory in the eyes of his peers and on occasion even leading to suicide.[21]

Codes of behaviour in the event of arrest and interrogation varied from one Zionist organization to the other. Among the members of Ha-Shomer ha-Tza'ir, the accepted practice was to completely deny one's membership in the movement until material evidence and testimonies presented during the interrogation made it impossible to refute. From that point on, the prisoner was expected to refuse to give evidence.[22] Members of Legal Hechalutz[23] were directed to protest their arrests on the grounds that their organization was not anti-Soviet and had not been banned in the USSR.[24]

Detainees usually spent the period from their arrest to the handing down of a sentence in an OGPU prison. The length of this preliminary detention ranged, as a rule, from three to six months. The investigation itself was usually carried out at the location of the arrest, but the materials prepared for the case were sent to GPU agents in Moscow, where the fate of the prisoner was decided, in his absence, by the Special Commission.

This procedure was not unique for the treatment of Zionists. Rather, it reflected the gradual emergence and consolidation of a correctional system for the new state. In November 1920 the People's Commissariat of Justice formally regulated the system of prison administration, the procedures and conditions for serving terms of imprisonment, issues of forced labour and education programmes for prisoners. The administration of places of imprisonment was entrusted to the Central Punitive Department of the People's Commissariat of Justice (Narkomiust). However, the correctional system of the OGPU continued to exist independently and did not fall under the jurisdiction of Narkomiust.[25] This was manifest in the establishment by the OGPU of the so-called 'special camps' (*spetslageria*) for political prisoners.

In keeping with the communist view that criminal activity was an abnormality resulting from class struggle, the tsarist system of incarceration was replaced by a system based on 'isolation' of anti-state elements. The term used initially for political prisons was *politizoliator*. On 15 June 1923 in the former Spaso-Evfimiev Monastery in Suzdal', a 'Special Concentration Camp' of the OGPU was formed and later renamed the Suzdal' *Politizoliator*. Likewise, the Solovetskii Forced Labour Camp for Special Purposes – better known as Solovki – was created in 1923 and two years later special prisons for political detainees began operating in Iaroslavl', Verkhne-Ural'sk, Cheliabinsk and Tobol'sk.[26]

Until January 1923, de facto jurisdiction over deportations of political detainees was in the hands the Special Bureau for Administrative Deportation of the Intelligentsia. From March 1923, it was transformed into the Technical Bureau on Administrative Deportation. In March 1924, the Politburo formally granted the OGPU the right to incarcerate deportees in camps for a period of up to three years.[27] A Special Commission (OSO) of the security police was later formed and given extra-judicial power to deport, exile and imprison in a concentration camp or a political prison, any individual implicated in anti-Soviet activity.

The OSO consisted of three members of the OGPU collegium (one of whom was the supervising prosecutor). The first Commission was comprised of Viacheslav Menzhinskii, Genrikh Iagoda and Gleb Bokii.[28] The institution survived until 1934, although its power diminished over time as a result of a process of decentralization within the OGPU and the increasing role of the evolving judicial organs to which Commissariat of Internal Affairs (NKVD) cases were being referred.[29]

The OSO received evaluations and recommendations for sentencing from local departments of the OGPU throughout the country. It met every 7–10 days to decide cases and then turned them over to the organs of the OGPU for implementation. The accused had no rights to be present or to offer any kind of defence.[30] The OSO decided on the type of punishment (*politizoliator*, exile), length of the sentence and the region to which a prisoner would be exiled. The specific site at which the exile would serve out his term was determined by the regional organs of the OGPU. Reaching the place of exile required the convict to endure the so-called 'convoy' (*etap*).

It was the OSO and its predecessors that handed down sentences of incarceration in political prisons or exile to Siberia, Turkestan, the Urals and to other remote regions to many convicted Zionists. Zionists arrested in small cities and regional centres were held in the local prisons until they constituted a sizable enough group to be transferred to a larger city. From there the convicts were usually sent to Moscow, where in the Butyrskaia, Taganskaia and Krasnopresnenskaia prisons convoys were formed to set out in various directions.

Conditions in the political prisons varied. Leah Cherkasskaia left the following description of her experiences in the prison in Iaroslavl':

> The political prison was in the old jail not far from the city. Before the revolution political prisoners had been incarcerated there on the way to labour camps [*katorga*]. Everyone had a separate cell. The cells were larger than those in Butyrskaia Prison [in Moscow] and, above all, they were clean. 'Lights out' came at 10:00 – just after evening exercise. We read a great deal and studied...I studied English.[31]

The harshest regime and worst physical and climactic conditions were at Solovki. The horrors of the Solovki camp were notorious even outside the USSR. Many who could not endure the harsh conditions fell ill and some perished. Seven members of the TzS were among those held there. In 1925, under pressure from the international community, all political prisoners held in Solovki, including Zionists, were transferred to other political prisons.[32]

The difficult conditions notwithstanding, the Bolshevik authorities continued to recognize the special status of 'political prisoners' until the beginning of the Great Terror. So-called 'politicals' were kept apart from criminals, and at least on paper, they had a number of rights, which were the legacy of the pre-revolutionary penal system. These included special food rations, cigarettes, rights to exercise, respectful forms of address by guards, access to newspapers, magazines and books, conjugal rights and certain privileges with respect to representation *vis-à-vis* prison officials.[33] Encouraged by the Political Red Cross (Pompolit), the prisoners defended these rights through hunger strikes or disobedience.

Prisoners did not always serve out their full sentences. Some were released after serving partial sentences and transferred to the status of exiles. The years in exile were often commuted to residency restrictions in cities or regions referred to as 'minus' zones. This meant, in practice, that the prisoners were forbidden to settle in Moscow, Leningrad, Ukraine, Belorussia or Crimea. They were obliged to report regularly (as often as several times a week) to the local department of the OGPU. However, these residency restrictions did not deprive one of the right to work or study, and enabled limited freedom of movement.

There are known cases of Zionists who were exiled as early as 1922. One of the first was a member of the organization Hechaver, who was arrested in Moscow in the middle of 1922 and charged with membership in the Secretariat of the Central Committee of the Zionist Movement. He was exiled for a year to Kostroma. Two members of the TzS, who were arrested in autumn 1922 in Kremenchug were also sentenced to exile but escaped.[34]

There seems to have been little if any logic or fixed pattern in determining the length or nature of individual sentences, but an average exile consisted of two to three years in remote regions of Siberia, the Urals and Central Asia. Punishments seem to have been more severe for activists of the TzS, because of its alleged Menshevik leanings.

Until the beginning of the 1930s, Zionist political prisoners were permitted to carry on a frequent and virtually uninterrupted correspondence with other exiled friends, relatives and comrades, as well as with those who had managed to leave for Palestine. The letters were read by agents of the OGPU, but usually reached their destinations, and those which were preserved, together with diaries and subsequent memoirs, supply a wealth of information on conditions of life in exile.[35]

Most Zionist convicts were sent to remote villages and towns isolated from the outside world in regions of harsh and unfamiliar climate. Upon arrival, the guards turned the exiles over to the local office of the OGPU, where they were registered and given documents assigning them a place of residence. After the ordeals of arrest, prison, awaiting sentencing, and finally, the torturous journey, the arrival at the place of exile was seen by many as form of liberation. Izrail' Mador, a member of Ha-Shomer ha-Tza'ir youth movement described this initial sense of liberation:

> They gave me papers for political prisoners, forbidding me to cross the city limits, and I was released. They even said to me: 'You are free.' After the long months of prison, I felt free to walk through the streets, to buy bread, to drink water from the well. It is true that I realized my freedom was relative, but all the same it was a wonderful sensation.[36]

In fact, the exiles were at the mercy of the local OGPU officials, who could, at the slightest suspicion of a transgression, or for no reason at all, arrest and imprison them and reassign them to exile in another location.

Many of the exiles, after months in prison and in transit, arrived sick, emaciated and in urgent need of medical care which was unavailable in these remote locations.[37] Their financial situation was usually desperate. Exiles received a monthly allowance of 6 rubles and 25 kopeks, but even this small sum was often not paid on a regular basis. It was extremely difficult to find any work in the very small settlements to which the prisoners had been exiled. Trained professionals in certain fields sometimes managed to establish themselves, but most of the exiled Zionists were young, had little education and no professional training. Moreover, permission to work had to be granted by the local OGPU and was difficult to obtain. Even when these permits were granted, income earned through temporary unskilled labour was negligible. The Zionists lived, therefore, in extreme poverty. They were crowded together in horrendous conditions and suffered from malnutrition. One exile recalled how she and her comrades, in the village of Uil in Siberia, ended up living in a mud hut. They managed to obtain logs to make a floor on which they slept. In other areas exiles survived on rice porridge, supplemented, on rare occasions, by fermented mare's milk, cheese and melon.[38] Another recalled that exile, which began with a sense of freedom, quickly turned into a fierce struggle for survival.[39]

In many cases the new arrivals to the places of exile were assisted by their comrades, who had been convicted earlier. The Zionists organized communes, shared lodgings and financial resources.[40] The communes played an important social role. The cooperation, the sense of camaraderie and the common ideology raised their morale and helped them to overcome the separation from families and friends. For many of the Zionists in exile, study was a main preoccupation. They sometimes had access to local libraries, and made use of them to read literature as well as studying mathematics, physics, social sciences, foreign languages, and for acquiring practical skills that might be useful in later life. In remote villages where no communal libraries existed, they often created their own. Many attempted to study Hebrew, but were hindered by the unavailability of textbooks and study materials.[41]

It was not uncommon for Zionist exiles to establish contact with other political exiles (generally socialists) living in the area, with whom they found common ground for social interaction and political debate. Moreover, many of the older exiled socialists were Jews, who were willing to share their experience and knowledge.

Despite the harsh climatic conditions, the poverty and the relentless and intrusive control of the local OGPU, for many, exile was tolerable, in particular because it was seen as merely a transitional episode on the road to Palestine. Certain places, where a relatively large number of Zionists were serving sentences, became nuclei for coordinating communication and assistance. In the small town of Poltoratsk, for example, as many as 15–25 Zionists served out their sentences at any given time. They formed a collective which maintained active contacts with Zionists sent to nearby villages in Turkmeniia.[42]

Many exiles eventually realized their dreams through the process of *zamena* – the substitution of their sentences of exile within the USSR, with exile to Palestine. Their experience of solidarity in the face of adversity in exile seems to have served them well in the communal life which they built there.

Substitution

The decision by Soviet authorities to allow sentences of internal exile to be substituted for emigration to Palestine set in motion a ten-year 'exodus' of convicted Zionists to Palestine. Yet, the evolution of this policy is somewhat shrouded in mystery. Deportation as a form of punishment was used by the GPU in the early 1920s in an unsystematic fashion.

Fedor Dan and a group of ten Mensheviks were deported abroad when they refused to accept their sentence of internal exile in January 1922.[43] In April 1922 the GPU offered deported members of the All-Russian Committee to Aid the Starving, publicist Ekaterina Kuskova and her husband Sergei Prokopovich, the option of substituting the sentence of internal exile which they were serving, with expulsion from the country.[44] Deportation from the country as a legitimate form of punishment for serious offences was formalized at the end of the Civil War in the Criminal Code which was passed in June 1922 – just in time for a major show trial of Socialist Revolutionaries (SRs) which took place from 8 June to 7 August 1922. Among Lenin's comments on the draft of the Criminal Code which he received in mid-May from the People's Commissar for Justice D.I. Kurskii. was an addendum to the Articles which concerned the death penalty. He suggested the following: 'Add the right, by decision of the Presidium of the All-Russian Central Executive Committee, to commute the death sentence to deportation (for a term or for life).'[45] In a note to Kurskii in the margin he wrote:

> I think the application of the death sentence should be extended (commutable to deportation)... to all forms of activity by the Mensheviks, SRs and so on to be formulated so as to identify these acts with those of the *international bourgeoisie* [Lenin's emphasis] and their struggle against us...

In a letter to Kurskii written two days later, Lenin offered a more precise formulation:

> Propaganda or agitation, or membership of, or assistance given to organizations the object of which (propaganda and agitation) is to assist that section of the international bourgeoisie which refuses to recognize the rights of the communist system of ownership that is superseding capitalism, and is striving to overthrow that system by violence, either by means of foreign intervention or blockade, or by espionage, financing the press, and similar means, is an offence punishable by death, which, if mitigating circumstance are proved, may be commuted to deprivation of liberty, or deportation.[46]

At the time Lenin wrote these words, he was already planning the expulsion from the country of intellectuals who were aiding the counterrevolution. He warned Dzerzhinskii that measures 'on the question of exiling abroad writers and professors who aid the counterrevolution' must be carefully prepared in order to avoid committing 'stupidities'. 'We must arrange the business in such a way as to catch these "military spies" and keep on catching them constantly and systematically and exiling them abroad'.[47] A Special Bureau for Administrative Deportation of the Intelligentsia was established and between 1922 and 1924 a number of decisions that stipulated permanent exile outside the borders of the

USSR were taken.[48] The most spectacular case occurred in September 1922 when a group of more than 100 cultural figures were forcibly deported from Russia to Germany.[49] They included well-known figures such as the philosophers Nikolai Berdiaev and Lev Karsavin, historian Sergei Mel'gunov and the literary critic and translator Iulii Aikhenval'd. Early in 1923 additional small groups were forcibly exiled. However, the damage which the deportation of leading figures in the arts and science caused to certain institutions and the enthusiasm with which Russian émigré communities in Europe received these 'outcasts' deterred extensive use of deportation.[50]

The notion of offering arrested Zionists the option of exile to Palestine was apparently the brainchild of renowned pianist, Professor David Shor.[51] The role played by him is described in detail in the Introduction to this volume. The policy of substitution may also have been influenced by considerable pressure applied by international Jewish organizations and the Jewish communities of Europe to end persecution of Zionists.[52] Bolstered by Dzerzhinskii's view that persecuting Zionists could cause more harm than good and by his conviction that 'the Zionists have great influence both in Poland and in America',[53] substitution was seen as a means to project a more positive image and win over influential Jewish communities abroad at a time when the new Soviet State was struggling for international recognition and the establishment of diplomatic relations with the West. In the process, the OGPU hoped that by deporting the leaders and activists it could weaken the activity of the Zionist movement in the USSR.[54]

Thus, beginning in 1924 and through the early 1930s, a large number of Zionists arrested throughout Russia were given the opportunity to substitute other forms of punishment with deportation to Palestine. In the early years, the OGPU would sometimes offer a detained Zionist deportation to Palestine as an alternative to internal exile in the course of an investigation.[55] In addition to the usual sentences of 'incarceration' and 'exile', the documents of the OSO now contained anomalous-looking verdicts which read: 'Palestine', 'Permission granted for Palestine', 'Palestine in the event of appeal', 'Siberia for three years or Palestine on request' and 'Urals for two years with substitution by Palestine'.[56]

A case which prefigured the wave of substitutions for Palestine occurred on 11 August 1923. Certain members of Tze'irei Tziyon who had been sentenced to imprisonment in 1922 were informed that their sentences were being substituted for deportation beyond the borders of the USSR. While waiting for their visas they were sent first to Kharkov, then to Moscow and Vladimir. On 21 January 1924, they left the Soviet Union for Germany.[57] When they arrived in Berlin, they carried out a campaign of organizing lectures and interviews in order to call attention to repression of the Zionist movement in the USSR. As a result the Soviet authorities revised their policy and stopped all deportations of Zionists through Europe.[58] From that time, 'substitution' prisoners had to travel directly to Palestine by sea from Odessa. The first deportees who had opted for substitution arrived in the port of Jaffa on the *Novorossiisk* steamship in July 1924.

Once substitution was approved, a convicted Zionist was either kept in prison or released and forbidden to travel while waiting for a passport and visa. Upon obtaining the relevant documents he was sent to Odessa and from there set sail for Palestine. As the numbers of requests increased the procedure was streamlined; less time was spent in prison – usually not more than 2 months,[59] although there were cases in which detainees who had received the substitution, were kept in prison for 6–7 months.[60]

In September 1925 the procedure was revised. The OGPU reviewed the 'punishment measures' applied up to that point and decided to eliminate 'direct' deportation of arrested Zionists. According to the new procedure, Zionists who had been detained by the OGPU had to be sentenced first to imprisonment or exile. Only at a later stage could the sentence be commuted to substitution.[61] In the overwhelming majority of cases, substitution was granted to those who were in internal exile. In 1931, when substitution requests of imprisoned Zionists Esfir' Krasnogorskaia and Boris Gal'perin were turned down by the OSO, they were informed that: 'Inmates in political prisons can not count on receiving permission to depart for Palestine', and 'only exile or residence limitations (minuses) can be replaced with permission to depart for Palestine. Deportation to Palestine can not be substituted for a sentence in a political prison'.[62] Nonetheless, some applications for substitution of imprisoned Zionists were approved.[63]

The impetus for reassessing and modifying the procedure came largely from the OGPU's realization that many Zionists were using substitution as an expedient means to get to Palestine. Indeed, the Chief of the Special Commission of the OGPU, T.M. Deribas, and head of the Special Commission's 4th Department Yakov Genkin argued in a memorandum to Dzerzhinskii of 29 May 1925, that the policy seemed, in fact, to encourage Zionist activity:

> When, up to the end of 1924 we were primarily deporting [Zionists] to Palestine, this became a major stimulus for intensifying national activities of the Zionists, since each one was sure that for his anti-Soviet activity he would receive the possibility to travel at public expense (of the Zionists or organizations sympathetic to them) to Palestine without paying the price for the crime he had committed.[64]

With this in mind the GPU sought to formulate more rigorous guidelines covering the substitution for exile to Palestine and specifying circumstances under which it should be denied. Based on past experience, it was decided that in general the activists, members of the Zionist Central Committees and Provincial Committees, those found producing and in possession of allegedly anti-Soviet literature would not be permitted to substitute exile with departure for Palestine, whereas the less active element would be given this option. The TzS was singled out in particular. Its members were to be sentenced only to internal exile and to imprisonment in concentration camps, and denied substitution because of the party's Menshevik leanings.[65]

Dzerzhinskii continued to have reservations regarding the efficacy of a more severe approach, and his protest is recorded at the bottom of the document:

> ...such broad persecution of Zionists, especially in the border regions, will not benefit us...in principle, we could be friends with the Zionists. This question needs to be studied and brought before the Politburo.[66] The Zionists have great influence both in Poland and in America. Why have them as our enemies?[67]

Some Zionist leaders sought to stem the tide of substitution, which they justifiably viewed as an attempt on the part of the authorities to drain the rank and file of the movement and in so doing paralyze its efforts. To ensure that they did not lose their most active and responsible members, some of the Zionist organizations required approval of a party committee, before an arrested member could apply for substitution. Another concern was that the prospects of emigration might draw opportunistic elements into their movements.[68]

It is difficult to calculate the total number of Zionists allowed to leave for Palestine in the period from 1924 to 1934. An assessment based on information from a variety of Zionist sources appears in Chapter 3 of this volume entitled 'Into Palestine'. What is clear is that the number of permits declined perceptibly after 1931 and by 1935, Zionist prisoners were being discouraged by Pompolit from initiating substitution as is evident from the words of one letter, 'there is no chance it will be approved'.[69]

The maze of bureaucracy and technical difficulties involved in applying for substitution and once it was received, in arranging and overcoming the obstacles involved in the actual departure, were overwhelming. Zionist political prisoners began to rely on the assistance of Ekaterina Peshkova and to forward their petitions to the OSO via the organization providing aid to political prisoners and exiles known as Pompolit.

Pompolit

Organizations aimed at assisting political prisoners had existed in Russia even in pre-revolutionary times. The first were created in St Petersburg as early as the mid-1870s with the participation of leading members of the Peoples' Will (Narodnaia Volia) L.I. Kornilova-Serdiukova, L.V. Sinegub and Vera Figner. The earliest official organization was the Red Cross Society of the People's Will – unofficially known as the Political Red Cross. Created by Iurii Bogdanovich and I.V. Kaliuzhnyi, it was funded largely by donations of sympathizers from among the liberal intelligentsia. Even after the obliteration of the People's Will, in the wake of the assassination of Alexander II in 1881, it continued to exist as a non-partisan organization under the name, Society for Aid to Political Exiles and Prisoners. Following the 1905 revolution, when many representatives of socialist parties were arrested and imprisoned, the number of organizations and societies providing aid to political prisoners increased. *Inter alia*, these included the

Krakow Union for Aid to Prisoners and the Committee for Aid to Political Prisoners in Russia.[70]

Aid to Political Prisoners (Pomoshch' politicheskim zakliuchennym) or Pompolit was established, de facto, in two documents dated 11 November 1922 and signed by the Deputy Chairman of the GPU, Iosif S. Unshlikht. It is noteworthy that these documents empowered E.P. Peshkova personally. She was given a mandate to 'render assistance to political prisoners',[71] but Pompolit itself was only indirectly referred to when permission was granted to stamp correspondence with the imprint, 'E.P. Peshkova, Aid to Political Prisoners'. An additional document, bearing the same date, formally transferred the offices of the former Moscow Political Red Cross and all its contents to Peshkova for her work in aiding prisoners and their families.[72] The office in question comprised a converted two-room apartment on the top floor of a residential building which was registered to the GPU. It was located next to the Lubianka, GPU headquarters in the heart of Moscow. Most strikingly, Peshkova was given direct access to the secret police in conducting her affairs. She was authorized to make oral or written personal appeals to the empowered representative of the GPU, to receive responses to her inquiries and requests and to receive communications from political prisoners via the investigative bodies of the GPU. The organs of the GPU were to cooperate with Peshkova 'by reviewing her petitions as matters of first priority'.

Thus, it would appear that the status of Pompolit wholly reflected the status of Peshkova herself. Ekaterina Pavlovna Peshkova (1876–1965), who headed Pompolit throughout its existence was born Ekaterina Volzhina, to a poor but aristocratic family in the Ukraine. Peshkova graduated from gymnasium in 1895 with a gold medal and began work as a proof reader at the newspaper *Samarskaia gazeta*. There she met Maxim Gorky (pseudonym of Aleksei Peshkov). After a very brief courtship, they were married in August 1896. They moved to the Crimea, where their son was born. In 1906 they were divorced but maintained extremely close relations; Peshkova, for example, continued in subsequent years to manage Gorky's personal finances.[73]

Gorky was an outspoken public figure of immense stature – both before and after the revolution. Although his status never gave him complete immunity from political persecution (he sat in prison and lived in self-imposed exile) the authorities treated him with caution. He fought for a number of social causes, condemned persecution of Jews (in 1915 even established the Russian Society of the Life of the Jews which fought against persecution of the Jewish community) and aided the SR and Social Democratic Labour Party. He was critical of Lenin and Trotsky, whom he once referred to as the 'Napoleons of socialism'. During the Civil War he reached a *modus vivendi* with the Bolsheviks, but his continued outspokenness led Lenin to pressure him to leave the country in 1921. He returned to the USSR in 1928 and continued to fight for humanitarian causes. During the famine of 1921, Gorky made international public appeals for aid. In 1922 he campaigned against the death sentence given to twelve leading SRs and throughout his life intervened on behalf of political prisoners.

Peshkova moved to Italy with her son in December 1906 and later to Paris – at the time the heart of Russian revolutionary émigré activity. She remained in Europe until September 1914. From 1908 to 1912 she was employed at the Emigré Fund (Immigrantskaia Kassa) in Paris which raised money for Russian political refugees.[74] In Paris she joined the Socialist Revolutionary Party. The precise date on which she became a member is difficult to discern, but this fact was noted by Gorky in his correspondence of 1908 and it is known that from 1907 she began working as an editor at the SR newspaper, *Znamia truda*. During the 1917 Revolution she was a member of the SR Central Committee.[75] In 1914 Peshkova returned to Russia and during WWI worked in several philanthropic organizations. She was the chairperson of the Committee of Assistance to Children – an organization to aid children and orphans abandoned at the front, which she established in 1915 together with the well-known lawyer E.N. Sakharov.[76] At the same time, she was active in the underground association known as The Circle to Aid Prisoners and Exiles (Kruzhok pomoshchi katorge i ssylke). In March 1917, following the February Revolution when tsarist political prisoners had already been released, the Circle became defunct but was soon reconstituted as the Society to Aid Prisoners and Exiles (Obshchestvo pomoshchi katorge i ssylke) and Peshkova continued to play a central role there. In March 1917, she travelled to St Petersburg on behalf of the Society in order to establish direct contact with the branch run by Vera Figner.[77]

The 1917 revolutions changed the entire political structure of the state. Tsarist political prisoners were freed, but were soon replaced with the enemies of the 'new socialist order'. Peshkova came to the aid of these inmates, some of whom now included friends, acquaintances and comrades-in-arms. In 1918 an organization known as the Moscow Political Red Cross was established and Peshkova became a leading activist there. In 1919 she was permitted by Dzerzhinskii to visit 'all prisons and other sites of incarceration of detainees [in Russia] and to interview them...'.[78]

In autumn 1920, in the war raging between the Bolsheviks and Poland, Peshkova became a representative of the Polish Red Cross and aided in its efforts to assist Polish refugees and prisoners on Soviet territory. This assistance was formalized with the conclusion of a peace treaty on 18 March 1921, which stipulated repatriation of Russians and Ukrainians on the one hand, and Poles on the other.[79] In her capacity as a representative of the Polish Red Cross, Peshkova was involved in the repatriation process and travelled to Siberia in September 1921 and in April 1922 to Arkhangel'sk to compile lists of prisoners and check on the status of those interned in camps there.

Although Peshkova officially continued to represent the interests of the Polish Red Cross up to 1932, her main efforts were increasingly directed towards providing aid to Russian political prisoners and exiles. As political arrests became more widespread, she began using her personal influence to intervene on behalf of some of the victims.

Gorky and Peshkova's interventions undoubtedly exploited the former's special status in the early years of the revolution. Thus Peshkova was often able to channel her appeals to Lenin through his sister, Maria Ul'ianova. Based on their reminiscences, Gorky and Peshkova's acquaintance with Feliks Dzerzhinskii began in 1909–1910, but was enhanced after the revolution. Dzerzhinskii is best known as the ruthless organizer and first chairman of the Extraordinary Committee for Combating Counterrevolution and Sabotage – the Cheka, but he was a peculiar character full of contradictions. He repeatedly petitioned the Council of Commissars for abolition of the death penalty, while presiding over shootings of huge numbers of people.[80] He earned the title 'Iron Feliks', but was also known as the 'friend of children' for his work with war orphans. Gorky wrote of Dzerzhinskii:

> In 1918–21, I came to know him closely, I spoke to him on several occasions concerning very sensitive issues, and often requested of him various interventions, and thanks to his sincere sympathy and sense of justice, much good was done.[81]

In her work in the Red Cross and in the various organizations to aid political prisoners, Peshkova had frequent contact with Dzerzhinskii and commented often on his 'responsiveness' and 'sense of justice'. When he died, she wrote of him in an open letter, 'there is no more wonderful, dearer man to everyone who knew him',[82] a comment which apparently got her into trouble with the Russian émigré communities.[83] Nonetheless, she seemed, generally, adept at negotiating her way successfully between circles of Bolshevik officials in the USSR and political émigrés abroad and was respected by both groups.

It is only in the context of her previous aid activities, her ties to the OGPU leadership, in particular to Dzerzhinskii, and her vast connections among the Bolshevik élite, made on her own and through Maxim Gorky, that one can understand her extraordinary status and the freedom of action and manoeuvrability within the GPU, which enabled Peshkova to operate Pompolit so effectively. She was given freedom to travel extensively at a time when the borders of the Soviet state were closed to all but a narrow circle of the power élite. She spent holidays with her son and with Gorky's family in Italy every year and nurtured contacts in Poland and Germany.

Although Pompolit had only a quasi-official status, its reputation grew rapidly. Hundreds of letters arrived every day to its offices from political prisoners and exiles with requests to resolve their specific problems and from relatives trying to ascertain the fate of their loved ones and ease their lot. The narrow corridor of the office was always crowded with people.[84] For the most part the employees filled purely administrative tasks: office manager, head of administration, records manager, accountants, typists, couriers, correspondents, freight handlers, etc.[85] Former political prisoners were often employed on a volunteer basis. There was a turnover of staff, but some of Pompolit's employees worked for the organization throughout its existence.[86] A sense of mission and camaraderie prevailed in the office. Staff members established a tradition of dining together.[87]

There are no indications in any of the financial documents that Peshkova ever received a salary or any compensation for her work, indicating, most likely, that it was carried out on a volunteer basis. Salaries were officially paid to as many as 15 or 16 staff members as employees of the state. Peshkova's right hand man, Mikhail L'vovich Vinaver, was a Polish Jew and engineer, who had also been well-acquainted with Dzerzhinskii prior to the revolution.

Pompolit had local offices in Kharkov, Leningrad and Poltava. In Kharkov it was headed by Lota Borisovna Sandomirskaia; in Poltava, V.G. Korolenko and his sister-in-law I.S. Ivanovskaia-Voloshenko directed the operations; and in Leningrad the Pompolit office was run by Vera Figner and later by S.P. Shvetsov. Figner, who was one of Russia's most vocal revolutionaries had been a member of the Executive Committee of the People's Will and had, herself, spent many years in tsarist prisons for her involvement in the assassination of Alexander II. She became something of a revolutionary icon and with her return from self-imposed exile from 1907 to 1917, she engaged primarily in aid and relief work.

The primary objective of Pompolit was to ease the fate of all those who were arrested and imprisoned for political activity. In the early 1920s the status of 'political prisoner' was recognized primarily for socialists. Though the mandate given to Peshkova in 1922 was probably intended for those individuals who were encompassed by the official definition of political prisoner, in practice she assisted numerous victims of repression including Orthodox clerics, Catholics, Tolstoyans and other non-socialists. Peshkova assisted vast numbers of Zionist political prisoners, and was often instrumental in their attempts to attain substitution of their sentences with deportation from the USSR. Many acknowledged her efforts in their memoirs with deep gratitude.[88] One of the people aided by Peshkova recalled the following:

> ...Especially for us, the Zionists, who were in those years still very inexperienced political prisoners, dispersed in all corners of the country, she was the main living contact with the outer world. The Zionists gave her particular difficulty. She had to become familiar with the various shades of Zionist trends, whose members were in the numerous Russian prisons. For her the whole problem of 'Palestine' was essentially new. In spite of this she cared for us with patience and great devotion. We would bother her very often, even if we knew in advance that her intervention would be in vain. I myself sinned against her not a little in this respect, and in the course of ten years I addressed 12 applications for 'exchange' [substitution] to her. Eleven times I received a negative answer, and every one of these negotiations was in the hands of Ekaterina Pavlovna who would also keep in touch with my family, at that time already in Palestine.[89]

The OGPU defined the aid which Peshkova could render to political prisoners in both material and non-material terms. She was entitled to raise money for her activity by organizing concerts and other events and to accept donations, to send food parcels and transfer funds to the places of the prisoners' incarceration. She was also given the right to serve as an intermediary for the transfer of parcels and money from family members.[90]

From the outset, Peshkova understood the importance of gathering and organizing as much information as possible on political prisoners – dates of sentences, terms, current whereabouts and even the addresses of relatives. The data reached Pompolit from four basic sources: from the political prisoners themselves, who often wrote letters to Peshkova even from transit prisons, and later from their place of imprisonment or exile; from relatives and other political prisoners who were in possession of information on the arrest and sentencing and at the same time were trying to find out about the subsequent fate or whereabouts of the convict;[91] from staff of the OGPU apparatus from whom Peshkova received data on the resolutions of the OSO and from the administrations of prisons and political wards where political prisoners were held.[92] The information was processed and compiled in a catalogue which kept the Pompolit staff updated and gave them efficient access to information on the status of prisoners and exiles.

Of particular significance was Pompolit's role as intermediary between political prisoners and exiles on the one hand and the apparatus of the OGPU on the other. Aided by her interventions, thousands of the most diverse petitions and requests from political prisoners addressed to the OGPU reached their destinations and Peshkova saw to it that responses were received from the appropriate bodies, as well. In Ekaterina Peshkova, the petitioner found an enterprising and experienced advocate. In the early years she and Vinaver often appealed directly to friends and acquaintances in the upper echelons of the OGPU, on behalf of political prisoners. Dzerzhinskii's response to one documented appeal on behalf of a political prisoner named Khovrin was an order to: 'Take on this case, and if there is no concrete evidence the prisoner can be released.'[93]

However, personal connections played only a limited role in Peshkova's monumental humanitarian efforts: she camped out on officials' doorsteps; wrote cover letters describing the prisoners' impoverished and desperate state; demanded the provision of medical care to those who needed it and registered complaints about abuses of power committed by OGPU agents in the localities. She gave a telling description of her own persistence:

> Though the work is becoming increasingly more difficult and I repeatedly 'knock on doors' where no one answers, or answer rudely, I continue doing my share and time and time again, I accomplish what I have set out to do.[94]

The very knowledge that someone in Moscow who had already provided assistance to many others, was sincerely trying to help, boosted morale for many of the prisoners. There are numerous documented cases in which her efforts to attain a measure of lenience or to ameliorate a prisoner's status met with success.

Another important function of Pompolit was the distribution of aid. Above and beyond moral and administrative support, the convicts were in dire need of basic necessities and, of course, money. Pompolit raised funds and collected clothing, medicine and food which was sent in parcels to prisoners and those in exile.

Peshkova was able to raise some of the money in the USSR (again, using Maxim Gorky and his connections) but most of the assistance came from abroad – primarily from the Russian émigré community. Charity organizations and funds also contributed. Contributions were either intended for general use, or sent by relatives and friends and earmarked for a particular person. However, Pompolit was never adequately funded and operated on a shoestring budget throughout its existence.

The amount of power and flexibility which Peshkova possessed and the access she had to the Soviet penitentiary system in the 1920s was indeed remarkable, however, in the years that followed, that freedom was substantially curtailed. The authorities progressively limited her visitation rights and freedom to interview political prisoners and transfer parcels.[95]

Stalin's gradual rise to power from 1925 saw the decline of the old Bolshevik élite with which Peshkova was closely associated. Dzerzhinskii's health deteriorated and he died of tuberculosis in 1926. The building of the new stalinist political apparatus introduced a harsher attitude towards any opposition and individuals condemned for political crimes, culminating in the Great Terror of 1936–1938. Pompolit found itself in a constant battle with increasingly powerful, repressive security services. Vinaver epitomized the situation, when he declared that under Dzerzhinskii, Pompolit had had the opportunity to help thousands, under Menzhinskii, hundreds and under Iagoda it was very hard to work at all.[96] But by 1925 even Dzerzhinskii, Peshkova's apparent patron, was being forced to cope with the changing circumstances. He instructed his personnel to limit her requests to visit political prisoners in the provinces: 'Do not give them general passes – give a separate one for each trip for a limited time period.'[97] Visitation rights were gradually reduced, coming to almost a complete halt by the end of the 1920s. At the same time, requests or petitions from the convicts were met increasingly with refusals.

In 1926 the employees of the Leningrad office of Pompolit were arrested, and in 1930, with the death of S.P. Shvetsov it was closed for good. The activity of the Poltava office ended completely in 1935 with the death of Ivanovskaia-Voloshenko.[98]

By the mid-1930s political prisoners were no longer accorded special status and the activity of Pompolit was severely curtailed, although Peshkova persisted in her personal efforts. Ignoring the obvious threat to her own safety, she continued to pursue the defence of prisoner's rights appealing to the NKVD and to its head, Nikolai Ezhov. Peshkova wrote dozens of letters, appeals and memoranda on the rights of political prisoners during this period. In them she drew attention to violations by the prison and camp authorities, complained that family members of convicts were not being informed about their sentences, that illegal methods were being employed in the course of interrogations, protested against torture inflicted by the wardens, entreated the authorities to agree to meetings and much more. She compiled letters from political prisoners testifying to blatant violations of the law and passed them on to the NKVD.[99] But her letters and appeals went unanswered and she was no longer able to meet with top ranking officials of the

NKVD. In 1937 they stopped accepting money and parcels from Pompolit altogether. Mikhail Vinaver was arrested, convicted and later disappeared without a trace.[100] On 15 June 1938 Peshkova was given forty-eight hours to bring the activities of Pompolit to an end.[101]

During the years of activity in Pompolit, Peshkova was responsible for helping hundreds, perhaps thousands of Zionists and their families. When convicted Zionists were faced with the option of substituting their sentences with exile to Palestine it was, yet again, Peshkova, who helped them negotiate the red tape. Requests for substitution were initiated by the prisoners themselves and addressed to the OSO. Following is one of the few surviving petitions in the archive of Pompolit. It is surely indicative of the format these petitions took:

> To the Board of the OGPU from exile D.M. Levitan
> Declaration
> On 25 January 1928 I was arrested in the city of Moscow by the Secret Operations Department of the OGPU for belonging to a Zionist grouping and by order of the OSO OGPU was exiled to the Urals for 3 years. My place of exile is Kargopol'skii Region, Shadrinskii District. I request that my exile be substituted with departure to Palestine. 29/10/1928.
> Signed,
> D. Levitan.[102]

Petitions were passed on to the OSO through the organs of the OGPU or through Pompolit and reviewed. The time it took to receive a response varied. In the 1920s, a decision was usually handed down 2–3 months after the receipt of a petition. In the 1930s this would often take half a year. When a negative response was given, the convict could file a new petition only after six months. In the case of a positive decision, confirmation had to be sent from the secretariat of the OSO. Parallel confirmation through the channels of the OGPU to the local office was sent in the form of a summary of the rulings of the OSO. It was this document which in fact provided the legal basis for permission to be granted to the convict to begin making arrangements for departure.

Quite often the actual dispatch of the decision of the OSO was delayed for various reasons, or the document itself was lost in the post. In such cases Peshkova intervened in order to ascertain the final ruling, to obtain official notification from the secretariat of the OSO and to expedite its dispatch to the location where the punishment was being served out. The approval for substitution did not usually bear a date of expiration, but in some cases the local organs of the OGPU placed limitations on its validity and threatened to revoke it if the exile failed to receive a passport within a certain period of time.[103] In principle, only the OSO could annul a substitution, but in practice local officials could trump up charges of transgressions and petition for its cancellation. In some cases, individuals who had been granted the substitution as early as 1925–1926 and for various reasons (illness, changes in family status, lack of the necessary funds for the departure, etc.) had not acted upon it immediately and wanted to make use of it several years later. In such cases, they were required to receive reconfirmation of the validity of the substitution from the OSO. If, for example, new charges had been levelled against the petitioner in the meantime, permission for departure was usually denied. But, even when this was not the case, the substitution received earlier was often considered insufficient and the exile was forced to re-petition the OSO for substitution.

Once substitution was granted, an exiled Zionist had to obtain a passport, an entry visa to Palestine and transit documents for a ship sailing from Odessa to Palestine. The so-called 'foreign travel passport', a document giving the right to leave the USSR, was issued to exiles at the local office of internal affairs after presentation of the necessary declaration, a registration card from the state shipping agency, Sovtorgflot, confirming that a place had been reserved on a ship sailing to Palestine, and a notice from the OGPU giving permission for the substitution. Very often the exiles resided in villages that were far removed from an administrative centre which could issue a passport, and were obliged to obtain permission from the local OGPU to travel to the nearest regional centre in order to procure the necessary documents. The procedure could take from one to three months and depended to a large extent on the speed with which the passports were prepared by the foreign section of the NKVD and prompt delivery by the local staff of the OGPU of the necessary notification of permission from the OSO. To complicate the situation even further, the passport was only valid for a limited period of time and fees were high, amounting in 1924 to 20 rubles[104] and increasing to between 30 and 72 rubles by 1926, depending on the place of issue and the social status of the convict.[105]

On 10 August 1926 new instructions of the People's Commissariat of Finance (Narkomfin) on foreign travel passports were published which established a scale of fees for the passport in accordance with the social and material status of the applicant.[106] Low paid workers were charged 55 rubles, while highly paid workers were charged twice as much. Ironically, those who were unemployed paid even higher fees – 220 rubles. The so-called *lishentsy*, a category to which numerous Jews belonged, were forced to pay the colossal sum of 330 rubles.[107] The overwhelming majority of exiled Zionists fell into the unemployed category. Only a few could produce the necessary certificate from a place of employment and receive the so-called 'discount' (*l'gotnye*) passports.[108] Since the monthly allowance received by a political exile in the 1930s was about 6 rubles, and the average pay of a worker, 60–80 rubles a month, this was a sum which was virtually unattainable without the assistance of relatives and friends or representatives of their organizations abroad.

The dire economic situation of the early 1930s led the Soviet leadership to seek any possible means to obtain hard currency. Among the measures designed to improve foreign currency reserves was the enactment of provisions for individuals in the West to 'ransom' relatives who wanted to leave the USSR. On 16 October 1932, a ruling of the Central Committee (Protocol 119) was issued which set the new passport tariff for Soviet citizens emigrating abroad at 500 rubles for labourers and 1,000 rubles for non-labourers. In accordance with existing law, a supplementary 10 per cent surcharge to benefit the Russian Red Cross was added. The price of the passport was calculated in gold rubles and had to be paid in hard currency through the Intourist travel office.[109] These sums were so inflated that on 16 July 1934 the Chairman of the Board of Intourist, V. Kurts, addressed a special memo to the Central Committee of the Communist Party in which he proposed lowering the established rates to 330 gold rubles out of purely economic considerations. He reiterated his belief that extorting money from relatives living outside the country in exchange for granting exit permits to Soviet citizens was still a viable source of hard currency for the state, but feared that the inflated passport prices would decrease the overall revenue gained from the enterprise.[110] Concurrent with the adoption of the previously mentioned ruling, it was decided that henceforth, political prisoners and exiles who had received permission to depart to Palestine, were to be considered free and therefore were to be required to pay for their passports in full and in hard currency.[111]

Notwithstanding these difficulties, until 1934 there was a steady stream of legal emigration of Soviet Jews to Palestine. Those who were able to pay for passports and transportation in hard currency, obtained permits more easily. As the number of permits which were issued dwindled in the 1930s, Intourist tried to continue to exploit the economic benefits of its Palestine connection. It came up with a plan involving the use of Soviet ships for the transport of Jewish emigrants from Germany, Poland and the Baltic to Palestine through the USSR. At the beginning of October 1934, a meeting took place in London between Intourist representatives, Dr Gelman of the Jewish Agency in London and the head of the Histadrut and member of the Zionist Executive, David Ben-Gurion. At that meeting it was suggested that Ben-Gurion travel to Moscow for further discussions on the question. In a letter dated 31 October 1934, Intourist Chairman Kurts tried to persuade People's Commissar for Foreign Affairs, Maksim Litvinov that transporting 7,000 people a year (approximately 300 emigrants per trip) to Palestine could bring income of about 250,000 gold rubles. In response Litvinov wrote to Kurts:

> I have received your letter... and I can inform you that from a political perspective there are no objections to [the idea of] providing transportation to Palestine nor to the arrival of representatives of relevant organizations. I doubt, however, given the overload in our transport system, the shortage of railroad cars, etc. that for the sum of 250 thousand gold rubles which you have proposed, it would be worth undertaking this. But this question, of course, is not for me to decide.[112]

However, the proposal to transport European Jews to Palestine via the USSR was not implemented and by 1934 the flow of emigrants from the Soviet Union to Palestine had stopped completely.[113]

In addition to a passport, the exile had to receive a visa issued by a British consular official in Russia. Receiving the latter was contingent upon possession of an Immigration Certificate from the Department of Immigration and Travel of the Government of Palestine. A detailed discussion of this process follows in Chapter 3: Into Palestine.

Obtaining the documentation from Soviet and British authorities was fraught with difficulties which could not have been overcome without the active engagement of Peshkova, who served on behalf of the prisoners as a liaison to the Soviet authorities, the British consulate in Moscow and the Immigration Centre of the Histadrut. From 1925 through 1934 her counterpart at the Immigration Centre was Chaim Halperin. Peshkova would receive from Halperin lists of certificates for entry into Palestine that had been issued by the Department of Immigration and Travel of the Government of Palestine and would then track down the recipients in prison, exile or restricted residency (minus zones) and inform them in writing, that permission had been granted. Administrative complications and delays often snarled the process. A convict would receive permission to enter Palestine prior to receiving official notification of substitution, or in some cases had not even submitted the required petitions. Sometimes the certificate expired prior to receipt of the other documents. If the delay was significant, it was necessary to officially extend the expiration date of the certificate.[114] In cases of minor delays, Peshkova tried to get around the problem by persuading the British Mission to issue a visa in spite of the expired certificate (during the period when visas were granted in Moscow), or by sending a letter to Sovtorgflot guaranteeing payment of a return ticket for anyone refused entry into Palestine because of an expired document (during the period in which there were no relations).[115] If a certificate had only recently expired, one was usually able to receive an entry visa in Jaffa. Many other obstacles emerged such as discrepancies between the names in the passports and on the entry certificates and the need to include wives, husbands, children and others in the visas.[116]

The collaboration between Peshkova and Halperin intensified towards the end of 1927, when, as a result of various obstacles, including the break in Anglo-Soviet relations, obtaining substitution became more difficult. They discussed the problems associated with bureaucracy and implementation of the substitution in the frequent correspondence between them. Providing financial aid and securing necessary documents were the central themes.[117] Over time they worked out a system of sharing joint lists containing all the necessary information on the convicted Zionists.[118] Peshkova sent information to Halperin regarding those who most urgently needed money, specifying the sums requested. Halperin scraped together funds, and periodically wired them to Peshkova both in rubles and in hard currency.[119] He stipulated to whom the money should be given and the amount that was to be allocated to each person.[120]

The final obstacle for those granted the substitution for departure were the transit documents. Purchasing tickets was, unquestionably, the easiest part of the procedure, as it did not require the involvement of the secret police organs. Deported Zionists were obliged to board the ships of Sovtorgflot.[121] The cheapest third class ticket for the Odessa – Jaffa route was 110 rubles. A ticket on the ship (without the listing of a specific date of departure) had to be reserved in advance since, in the tangle of Soviet bureaucracy, it was impossible to obtain a passport for travel abroad without a card confirming a reservation. It was, therefore, necessary to forward the travel bureau an advance payment in return for a receipt guaranteeing a place on the ship, conditional upon payment of the balance of the ticket price. Before 1927, the advance payment was 20 rubles, but later the sum was increased to 50 rubles.[122] On arrival in Odessa, the traveller had to go to the office of Sovtorgflot, pay the balance and receive a ticket on the next ship departing for Palestine.

Only when an exile possessed a passport and visa with a receipt in hand showing advance payment for the ticket, could he go to the local office of the OGPU where he was given permission to leave for Odessa in order to board the ship.[123] In some cases individuals who received the substitution found themselves in a situation in which they had been given a strict deadline for departure (connected, for example, with the expiration of their passports, the substitution ruling or entry permit), but still had not obtained a visa in their passports, a Turkish visa, or a certificate. Peshkova was able, on more than one occasion, to attain permission from the OGPU for departure without a passport (or certificate), promising that she would transfer the missing documents to Odessa to be collected by the recipient at the post office there.[124]

The only practical way of reaching Odessa at the time was by rail. Quite often Zionists departing to Palestine requested permission to stop off along the way to part from relatives and loved ones. Such requests were generally denied by the OGPU and the deportees were instructed to proceed to Odessa by the shortest possible route. Peshkova advised them that 'Requesting permission from the OGPU to stop of at home before leaving for Palestine is useless. They always refuse this.'[125] Upon arriving in Odessa, they were to sign in on the registry of the local OGPU office and report daily until departure.[126] When problems arose, Peshkova was just about the only hope for salvation.[127] The files of Pompolit contain copies of hundreds of telegrams which were sent to Sovtorgflot, to the local Odessa office of the OGPU and to departing passengers. Peshkova tirelessly explained inaccuracies, vouched for the identities of those about to embark when discrepancies in versions and spellings of names appeared in their documents, guaranteed payments for return travel for those who possessed expired visas or entry permits, urgently dispatched money and documents that were missing and much more.[128] Also recorded in the archive of Pompolit is evidence of the enormous gratitude felt by the Zionist deportees whom she helped.[129]

The route from Odessa to Palestine via Constantinople was serviced by several vessels.[130] Ships sailed approximately once in every two weeks. The voyage usually took from 10 to 14 days, after which the ship arrived in Jaffa or Haifa harbour. Together with the Jewish emigrants from Russia, residents of Palestine returning home and a small number of tourists, the Zionist exiles disembarked to begin a new life.

Notes

1 V.I. Lenin, 'Polozhenie Bunda v partii', *Polnoe sobranie sochinenii*, Vol. 8 (Moscow: Politizdat, 1958–1965), pp. 72–4.
2 Iu. Margolin, *Kak bylo likvidirovano sionistskoe dvizhenie v Sovetskoi Rossii* (Jerusalem, 1988), pp. 7–8.
3 See *Hechalutz be-Rusya – le-Toldot Hechalutz ha-Bilti Legali (Ha-Leumi Amlani)* (Tel Aviv: Mishlachat Chul shel Hechalutz ha-Bilti Legali be-Rusya, 1932), p. 170.
4 Central Zionist Archives (CZA) F30/f. 140/1.
5 *Hechalutz be-Rusya*, p. 170.
6 CZA F30/f. 40/1.
7 See Ar'ye Refa'eli (Tzentziper), *Ba-Ma'avak li-Ge'ula. Sefer ha-Tziyonut ha-Rusit mi-Mahapekhat 1917 ad Yameinu* (Tel Aviv: Dvir, 1956), pp. 62–4. *Izvestiia VTsIK* (104) 16 May 1920. Other incidents included arrests of Zionists in Zhitomir, Odessa, Samara and Ekaterinburg in 1920.
8 *Hechalutz be-Rusya*, p. 171.
9 See *Rassvet* 28, 22 October 1922, p. 20; *Rassvet* 31, 12 November 1922, p. 20; *Rassvet* 33, 26 November 1922, p. 21 and *passim*. CZA F30/f. 140/1.
10 For details see Benjamin West (ed.), *Struggles of a Generation: The Jews Under Soviet Rule* (Tel Aviv: Massadah, 1959), pp. 169–70.
11 On 9 February 1922, the All-Russian Extraordinary Commission for Fighting Counter-revolution and Sabotage (VChK) was reorganized into the GPU, which existed until November 1923, when it was transformed into the United State Political Directorate (OGPU). In 1934, the OGPU became part of the People's Commissariat of Internal Affairs (NKVD).
12 See RGASPI (formerly RTsKhIDNI), f. 17, op. 84, d. 643, ll. 4–6.
13 For details see *Hechalutz be-Rusya*, pp. 171–2; Refa'eli, *Ba-Ma'avak li-Ge'ula*, pp. 138, 141–3; Yitzchak Ma'or, *Ha-Tnu'a ha-Tziyonit be-Rusya me-Reshita ve-ad Yameinu* (Jerusalem: Ha-Sifriya ha-Tziyonit, 1986), p. 534; Yehuda Erez (ed.), *Sefer TzS. Le-korot ha-Miflaga ha-Tziyonit-Sotzialistit u-Brit No'ar TzS be-Brit ha-Mo'atzot* (Tel Aviv: Am Oved, 1963), pp. 541–3, 535–7, 546, and *passim*; West, *Struggles of a Generation*, p. 170 and *passim*.
14 See Document 7 in this volume: F.E. Dzerzhinskii to V.R. Menzhinskii.
15 Aleksandr Kokurin and Nikita Petrov, 'GPU–OGPU (1922–1928)', *Svobodnaia mysl'* 7 (1998), pp. 113–20.
16 Based on accounts of survivors such as: Avraham Itai, *Korot Ha-Shomer ha-Tza'ir be-S.S.S.R. No'ar Tzofi Chalutzi-NeTzaCh* (Jerusalem: Ha-Aguda le-Cheker Tfutzot Israel, 1981), pp. 243–5, 287–90, 295–6; *Magen: Kovetz mukdash le-she'elot ha-Tziyonut ha-nirdefet be-Rusya ha-sovyetit* (Tel Aviv: Magen, 1931), pp. 4–11; Erez, *Sefer TzS*, pp. 389–90, 403–5, 416–19; 421–2, 423–6, 435–7; 541–3; Vera Kaplan, *K tebe dusha izdaleka...* (Tel Aviv: Pilies Studio, 1999), pp. 101–6.
17 See Refa'eli, *Ba-Ma'avak li-Ge'ula*, pp. 181–2.
18 Ibid.
19 Ibid., pp. 403–5.
20 See, for example, the memoirs of Moshe Ashuakh (Krymker) and Bracha R. in Erez, *Sefer TzS*, pp. 404–11, 416–19, and *passim*.
21 Margolin, *Kak bylo likvidirovano sionistskoe dvizhenie v Sovetskoi Rossii*, p. 13.
22 Itai, *Korot Ha-Shomer ha-Tza'ir be-S.S.S.R.*, pp. 287–90.
23 See the Introduction to this volume.
24 Dan Pines, *Hechalutz be-Kur ha-Mahapekha – Korot Histadrut Hechalutz be-Rusya* (Tel Aviv: Davar, 1938), pp. 276–7.
25 *Sobranie uzakonennii i rasporiazhenii Rabochego i Krest'ianskogo Pravitel'stva*, vyp. 23–44 (Moscow: 1921), pp. 129–46.
26 See Kokurin and Petrov, 'GPU–OGPU (1922–1928)', pp. 117, 119, 122.

27 See 'Protokol zasedaniia Politbiuro', 76, punkt 18, RGASPI, f. 17, op. 3, d. 24, l. 6.
28 OGPU order no. 250, dated 12 June 1924 in Kokurin and Petrov, 'GPU–OGPU (1922–1928)', p. 121.
29 See Kokurin and Petrov, 'GPU-OGPU (1922–1928)', pp. 120–1; Aleksandr Kokurin and Nikita Petrov, 'OGPU (1929–1934)' *Svobodnaia mysl'*, 8 (1998) p. 109.
30 See GARF, f. 8409, op. 1, dd. 56, 103, 208, 293, 410, 'Spisok postanovlenii Osobogo Soveshchaniia pri OGPU' for various years.
31 The memoirs of Leah Cherkasskaia in Erez, *Sefer TzS*, pp. 403–5.
32 See Ar'ye Refa'eli (Tzentziper), *Eser Shnot Redifot. Megilat ha-Gzeirot shel ha-Tnu'a ha-Tziyonit be-Rusya ha-Sovyetit* (Tel Aviv: Ha-Ve'ada ha-Historit shel Brit Kibbutz Galuyot, 1930), pp. 171–2.
33 A.I. Solzhenitsyn, *Arkhipelag GULAG* (Moscow: Inkom NV, 1991), Vol. 1, pp. 328–9.
34 Refa'eli, *Eser shnot redifot*, pp. 111–13.
35 See for example, Binyamin Vest (West), *Bein Ye'ush le-Tikva. Mikhtavim shel Asirei Tziyon be-Rusya ha-Sovyetit* (Tel Aviv: Reshafim, 1973), The bulk of the correspondence is preserved in the CZA F30/ff. 146–9.
36 Itai, *Korot Ha-Shomer ha-Tza'ir be-S.S.S.R.*, p. 310.
37 See, for example, the memoirs of Hanna Nudel'man in Erez, *Sefer TzS*, pp. 421–2; the letters of Matil'da Galperina, Sima Rubina and other in Binyamin Vest, *Naftulei Dor – Korot Tnu'at ha-Avoda ha-Tziyonit Tze'irei Tziyon Hitachdut be-Rusya ha-Sovyetit* (Tel Aviv: Delegation of Tze'irei Tziyon-Hitachdut in Russia, 1956), pp. 97, 203–5. and *passim*.
38 See Itai, *Korot Ha-Shomer ha-Tza'ir be-S.S.S.R.*, pp. 311–12; See Vest, *Naftulei Dor*. pp. 82–3.
39 Kaplan, *K tebe dusha izdaleka...*, pp. 118–21.
40 See, for example Erez, *Sefer TzS*, pp. 415, 423–6, 435–7; Itai, *Korot Ha-Shomer ha-Tza'ir be-S.S.S.R.*, pp. 310–14, 321–6; Vest, *Bein Ye'ush le-Tikva*, pp. 44, 51; and *passim*.
41 Vest, *Bein Ye'ush le-Tikva*, pp. 82–3, 149–50; Erez, *Sefer TzS*, p. 415.
42 A clear example of this activity can be seen in the very lively correspondence between groups of Zionists exiled to Poltoratsk and Pompolit. GARF, f. 8409, op. 1, d. 161, ll. 16, 30, 202, 300.
43 Stuart Finkel, 'Purging the Public Intellectual: The 1922 Expulsions from Soviet Russia'. *The Russian Review 62* (October 2003), p. 601. See also the Introduction to this volume.
44 Finkel, 'Purging the Public Intellectual', pp. 594, 601–2.
45 V.I. Lenin, 'Addendum to the Draft Preamble Criminal Code of the RSFSR', 15 May 1922, http://www2.cddc.vt.edu/marxists/archive/lenin/works/1922/may/15b.htm.
46 V.I. Lenin, 'Letter to D.I. Kurskii', 17 May 1922, http://www2.cddc.vt.edu/marxists/archive/lenin/works/1922/may/17.htm.
47 Discussions of this can be found in A.I. Solzhenitsyn, *The Gulag Archipelago* (NY: Harper and Row, 1974), p. 372 and in Finkel, 'Purging the Public Intellectual', p. 602.
48 GARF, f. 8409, op. 1, d. 56, ll. 62ob., 111, 150, 207 and others.
49 E.I. Pivovar (ed.), *Rossiia v izgnanii. Sud'ba rossiiskikh emigrantov za rubezhom* (Moscow: UVN PAN, 1999) p. 385. For a detailed discussion see Finkel, 'Purging the Public Intellectual'.
50 See, especially, Finkel, 'Purging the Public Intellectual', pp. 608–10 and Solzhenitsyn, *The Gulag Archipelago*, p. 372.
51 See Pines, *Hechalutz be-Kur ha-Mahapekha*, p. 273; Ma'or, *Ha-Tnu'a ha-Tziyonit be-Rusya*, pp. 535–6; Refa'eli, *Ba-Ma'avak li-Ge'ula*, p. 124; RGASPI, f. 17, op. 84. d. 485, l. 8; CZA F30/f. 909 (An article devoted to the seventy-fifth birthday of David Shor, *Davar*, 13 January 1939).
52 Ma'or, *Ha-Tnu'a ha-Tziyonit be-Rusya*, p. 535.
53 See Document 7: F.E. Dzerzhinskii to V.R. Menzhinskii and Document 8: Report of Special Department of OGPU to F.E. Dzerzhinskii, in this volume.

54 See RGASPI, f. 445, op. 1, d. 119, l. 3. Pines, *Hechalutz be-Kur ha-Mahapekha*, p. 276–7; Refa'eli, *Eser shnot redifot*, pp. 116–18.
55 Ibid.
56 GARF, f. 8409, op. 1, d. 56, ll. 28, 82, 84, 86–7, 93, 107ob., 117, 118.
57 For more information see *Hechalutz be-Rusya*, pp. 170–1; Refa'eli, *Ba-Ma'avak li-Ge'ula*, pp. 131–4.
58 Refa'eli, *Eser shnot redifot*, pp. 116–18; Pines, *Hechalutz be-Kur ha-Mahapekha*, p. 274.
59 See, for example, the responses of Ester Zhelezniak (144), Shai Zarubinskii (145), Ester Delegach (196), Mordechai Iampolskii (219) and others. CZA F30, ff. 123/1, 123/2, 124/1.
60 See the responses of Max Gorkin (147), Lidia Eiges (217) and others. CZA F30/ff. 123/1, 124/1.
61 RGASPI f. 76, op. 3, d. 26, l. 5.
62 GARF, f. 8409, op. 1, d. 604, l. 268; d. 682, l. 85.
63 See Refa'eli, *Eser shnot redifot*, pp. 255–6; Itai, *Korot Ha-Shomer ha-Tza'ir be-S.S.S.R.*, pp. 282–4 and *passim*.
64 See Document 8: Report of Special Department of OGPU to F.E. Dzerzhinskii, in this volume.
65 Ibid.
66 A thorough check of the protocols did not turn up any evidence that the issue was indeed ever raised in meetings of the Politburo.
67 See Document 8: Report of Special Department of OGPU to F.E. Dzerzhinskii, in this volume.
68 Refa'eli, *Eser shnot redifot*, pp. 116–18; Pines, *Hechalutz be-Kur ha-Mahapekha*, p. 276–7.
69 GARF, f. 8409, op. 1, d. 1414, registration number 2297 dated 4 July 1935 (pages not numbered); ibid., registration number 425 dated 22 October 1935 (pages not numbered); the files contain many analogous letters.
70 See V. N. Figner, *Komitet pomoshchi katorzhanam* (Paris, 1911).
71 GARF f. 8409, op. 1, d. 11, l. 1.
72 GARF f. 8409, op. 1, d. 11, l. 3.
73 Arkhiv A.M. Gor'kogo, Vol. 13, *Maksim Gor'kii i syn. Pis'ma. Vospominaniia* (Moscow: Nauka, 1971), pp. 17, 219; Anri Troiat, *Maksim Gor'kii* (Moscow: EKSMO, 2004), pp. 94–6.
74 Arkhiv A.M. Gor'kogo, Vol. 9, *Pis'ma k E.P. Peshkovoi, 1906–1932* (Moscow: Khudozhestvennaia literatura, 1966), p. 393.
75 Ibid., pp. 383–4.
76 Ibid., p. 365.
77 Ibid., p. 378.
78 From a certificate with Dzerzhinskii's signature dated 15 October 1919. GARF, f. 8409, op. 1, d. 1729, l. 9.
79 The peace treaty was signed in Riga on 18 March 1921. It was preceded by the 'Agreement on Repatriation', signed on 12 October 1920, which stipulated 'immediate repatriation of all hostages, civilian prisoners, interned persons, prisoners of war, refugees and emigrants'. See *Sobranie uzakonenii i rasporiazhenii Rabochego i Krest'ianskogo pravitel'stva*, vyp. 41–2 (Moscow: 1921), pp. 217–45.
80 Adam B. Ulam, *The Bolsheviks* (Toronto: Collier), 1985, p. 420.
81 Letter to Ia.S. Ganetskii, *Izvestiia*, 11 Aug 1926 – published on the day of Dzerzhinskii's death.
82 Aleksei M. Gor'kii, *Sobranie sochinenii*, Vol. 29 (Moscow: Gos. izdatel'stvo khudozhestvennoi literatury, 1949), p. 473.
83 Ibid., p. 409.
84 Haim Koserovski, 'Yekaterina Pavlovna Peshkova', in West (ed.), *Struggles of a Generation*, p. 184.

85 See GARF, f. 8409, op. 1, dd. 1452, 1633, 1693, 1694.
86 See A. Iu., Gorcheva, *Spiski E. P. Peshkovoi* (Moscow: 1997), p. 18 [n.p.].
87 GARF, f. 8409, op. 1, d. 1463, ll. 6–7, 19; d. 1464, ll. 5, 29–3.
88 See, for example, the memoirs of Moisei Gabin, Matil'da Gal'perina. Girsh (Tzvi) Vinogradskii, Ariel Pol'skii, and Izrail' Kaplan in Vest, *Bein Ye'ush le-Tikva*, pp. 67, 95–7, 113, 165–6, 198; Rachel Blechman in Kaplan, *K tebe dusha izdaleka...*, pp. 112–14. Koserovski, 'Yekaterina Pavlovna Peshkova', pp. 184–7.
89 Koserovski, 'Yekaterina Pavlovna Peshkova', p. 186.
90 GARF, f. 8409, op. 1, d. 11. l. 1.
91 Families were not informed of a prisoner's sentence. Very often it took months before some communication was received and until that point they had no knowledge of his or her fate.
92 GARF, f. 8409, op. 1, dd. 11–13.
93 RGASPI, f. 76, op. 3, d. 49, l. 42.
94 Quoted in Koserovski, 'Yekaterina Pavlovna Peshkova', p. 187.
95 GARF, f. 8409, op. 1, d. 1729, ll. 19, 27.
96 Gorcheva, *Spiski E. P. Peshkovoi*, p. 19. Dzerzhinskii served in the post of Chairman of the OGPU up to 1926. Viacheslav Menzhinskii was the Chairman of OGPU up to his death in 1934. Genrikh Iagoda headed the NKVD until September 1936. He was succeeded in the same year by Nikolai Ezhov.
97 To G.L. Gerson, assistant administrator of affairs (*upravdelami*) of the OGPU, RGASPI, f. 76, op. 3, d. 49, l. 42.
98 According to Solzhenitsyn, The Petrograd/Leningrad office which was headed by the Narodnik Shvetsov, 'adopted an intolerably impudent stance', mixed into political cases, tried to get support from such former inmates of the Schlusselburg prison as Novorusskii and helped not only socialists but also counterrevolutionaries and was thus closed down early. The Moscow office 'behaved itself' and thus survived until 1937. Solzhenitsyn, *The Gulag Archipelago*, p. 41.
99 See for instance, GARF, f. 8409, op. 1, d. 1557, l. 422; d. 1631, ll. 16, 25; d. 1636, l. 14.
100 Gorcheva, *Spiski E. P. Peshkovoi*, pp. 19–20.
101 Ibid., p. 43.
102 GARF, f. 8409, op. 1, d. 223, l. 199.
103 GARF, f. 8409, op. 1, d. 325, l. 236.
104 See, for example, CZA F30/ff. 123/1, 123/2.
105 To give a sense of how chaotic the price structure was, payment ranged from 30 to over 200 rubles, CZA F30/ff. 123/1, 123/2, 124.
106 *Pravda*, 11 August 1926.
107 See, for example, GARF, f. 8409, op. 1, d. 107, ll. 457, 178, 137; d. 381, l. 441. See Introduction to this volume regarding *lishentsy*.
108 See, for example, GARF, f. 8409, op. 1, d. 183-a, l. 248.
109 Arkhiv vneshnei politiki MID RF, f. 5 (Secretariat of Litvinov), op. 14, d. 3, papka 95, l. 104.
110 Ibid.
111 See, for example, the memoirs of Yakov Gorevoi (Gur Avi) in Vest, *Bein Ye'ush le-Tikva*, pp. 71–2.
112 Arkhiv vneshnei politiki MID RF, f. 5, op. 14, d. 3, papka 95, ll. 148–154, 156.
113 As a result in 1936 there were no runs from Odessa to Jaffa at all. In 1937 the service was restored but ships sailed less frequently. Archive of the Informational Centre of the Jewish Agency, Vols 34–9.
114 See, for example, GARF, f. 8409, op. 1, d. 199, l. 164; d. 2754, l. 93; d. 385, l. 202; d. 232, Correspondence with the Central Zionist Department in Palestine on the receipt of visas and the Right to Depart for Palestine.
115 See, for example, GARF, f. 8409, op. 1, d. 349, l. 258; d. 394, l. 317; d. 455, l. 246; d. 476, l. 175.

116 In some cases exiled Zionists managed to get married and even had children while waiting for their visas. Since a wife, as a rule, had the right to leave together with a spouse who had received permission, this sometimes led to marriages of convenience between comrades in the Zionist movements. See, for example, the memoirs of Lea Gol'dina in Erez, *Sefer TzS*, pp. 389–90.

117 See, for example, GARF, f. 8409, op. 1, d. 232, ll. 34, 34ob., 37, 41, 49, 73–4, 76; GARF, f. 8409, op. 1, d. 199, l. 164; d. 390, ll. 36, 113–14, 293; d. 445, l. 6. See also the memoirs of El'kan (Kuni) Bondar', Bunia Blat, Yakov Gorevoi, Vera Druian, Girsh Vinogradskii, Ariel' Pol'skii, and Izrael' Kaplan in Vest, *Bein Ye'ush le-Tikva*, pp. 40–2, 53, 71–2, 84–5, 113, 165–6, 197–8.

118 See, for example, GARF, f. 8409, op. 1, d. 232, ll. 12, 38–40, 63, 73; d. 521 (containing four pages of unnumbered lists); d. 553 containing seven pages of lists (unnumbered); and d. 696 containing thirteen pages of lists (unnumbered). Many were covered with Peshkova's handwritten notations: 'Visa received', 'Departed', 'Permission', 'Refusal'. 'Three years of political prison', from which it is apparent that they were in constant use.

119 GARF, f. 8409, op. 1, d. 232, l. 65, 68, 70, 73–4 and *passim*. Regarding sources of funding for the Zionist movement in the USSR see Chapter 3: Into Palestine, in this volume.

120 GARF, f. 8409, op. 1, d. 232, l. 34.

121 Formerly called Soiuzflot.

122 GARF, f. 8409, op. 1, d. 178, ll. 34, 36.

123 See, for example, GARF, f. 8409, op. 1, d. 200, ll. 189, 217; d. 350, l. 48; d. 368, l. 129.

124 See, for example, GARF, f. 8409, op. 1, d. 193, ll. 25, 29; d. 200, l. 192; d. 439, l. 43.

125 GARF, f. 8409, op. 1, d. 350, ll. 72, 138.

126 See the memoirs of Leah Cherkasskaia, in Erez, *Sefer TzS*, p. 405.

127 For example, GARF, f. 8409, op. 1, d. 193, ll. 76, 80, d. 422, l. 344, 348, d. 426, l. 49, d. 628, ll. 176, 181.

128 For example, GARF, f. 8409, op. 1, d. 426, l. 35; d. 35; d. 439, l. 263; d. 450, l. 208; d. 623, l. 225; d. 631, l. 77.

129 See, for example, GARF, f. 8409, op. 1, d. 200, l. 192 and d. 252, l. 200.

130 The route was serviced by vessels of the Black Sea shipping fleet: *Chicherin, Il'ich, Lenin, Novorossiisk, Tobol'sk, Sevastopol', Frantz Mehring* and *Krym*. In 1924 according to the data of the Immigration Department there were 3 voyages; in 1925, 27; in 1926, 23; in 1927, 20; in 1928, 25; in 1929, 24; in 1930, 27; in 1931, 25; in 1932, 21; in 1933, 22; in 1934, 26; in 1935, 20; in 1936, 0; in 1937, 10. Archive of the Informational Centre of the Jewish Agency, Vols 7–39.

3 Into Palestine
The Zionists and the British

Ziva Galili

The unusual arrangement allowing over a thousand Zionist convicts to leave Soviet Russia was not sufficient in itself to ensure their arrival in Palestine. Their immigration faced significant obstacles – political, administrative and financial – and required active support from several actors outside the Soviet Union. Without the cooperation of Great Britain, which had been entrusted by the League of Nations with a mandate to rule Palestine, the immigration from Soviet Russia could never have happened. Such cooperation was obtained partially through the efforts of Zionist lobbyists, especially the London Executive of the World Zionist Organization and the Palestine Zionist Executive in Jerusalem, who enjoyed access to British officials at all levels. These Zionist Executives, in turn, were under pressure from the Immigration Centre of the Histadrut (General Federation of Jewish Workers) and from representatives of the Zionist movements active in Soviet Russia, which demanded help for Zionist convicts and provided crucial information about their plight.

The interconnection among these players is apparent when we track the flow of papers carrying information and authorization for the immigration of Zionist convicts. The names of those arrested and those permitted to leave poured into the Immigration Centre from relatives, from the movements in Russia and their 'delegations' in Palestine and from the Committee to Aid Political Prisoners known as Pompolit. The Immigration Centre sent lists of arrested Zionists to the Immigration Department of the Palestine Zionist Executive (PZE), often providing information on each individual's age, place of residence or exile and political affiliation. The lists were reorganized and consolidated and then submitted by the PZE to the Immigration and Travel Department of the Government of Palestine (also called the Immigration or Permit Section). For most of the decade of substitution immigration, the decision to grant Immigration Certificates was made at that level, by the Controller of Immigration, but for a number of years (1927–1931) the lists of prospective immigrants from Russia required approval by the Home Office and Scotland Yard. Decisions authorizing certificates or rejecting individuals for immigration were conveyed in reverse order: from London to the Palestine Government, from the Controller of Immigration to the PZE's Immigration Department, and from there, finally, to the Immigration Centre, which informed prospective immigrants of the status of their requests through available channels (at first – the movements, later – Pompolit).

The story of how Zionist exiles were allowed to enter Palestine and assisted in travelling there has not been studied to date. Claims of a secret British understanding with the Soviet authorities compete with the more pervasive view of the British as intent on limiting Jewish immigration to Palestine. Some blame the Zionist Organization itself for neglecting the Zionist victims of Soviet persecution. 'The sins of the Zionist Executive are so numerous', announced a member of the Immigration Centre in autumn 1924, 'one can not count them'.[1] But as demonstrated by the documents used and reproduced in this volume – British government documents, documentation emanating from the Zionist Executives in London and Jerusalem, and the correspondence between the Immigration Centre, Hechalutz and other labour Zionist groups – there was widespread support for the Zionists persecuted in Soviet Russia. As always, the motives varied: there was the desire of the Soviet Zionist movements and the corresponding sectors of the Histadrut to increase the number of their political supporters in Palestine, the susceptibility of Zionist organizations to moral claims on behalf of imprisoned Zionists, and, finally, the anti-Soviet sentiment that underlay British tolerance towards victims of the Soviet regime. Beyond such motives, certain key individuals showed keen sympathy for these victims, or were moved by the belief that their immigration would be of special benefit to the Jewish settlement in Palestine.

In spite of the broad-based support, as we shall see, interactions among the organizations concerned with Zionist immigration from Soviet Russia were frequently strained by mutual mistrust. Support for the substitution immigration was also weakened by the resolve of the PZE as well as the British to prevent radical socialists from entering Palestine and the turn away from earlier pro-Zionist positions by many officials of the British Government in Palestine.

The Russian Zionist movements and labour politics in Palestine

Those members of the Zionist movements in Soviet Russia who made their way to Palestine were expected by comrades left behind to be their tireless advocates. Letters from exile and prison repeatedly urged them to do all in their power to arrange for the release and departure of the prisoners. These exhortations aside, the Russian Zionists in Palestine knew intimately the high expectations as well as the daily difficulties experienced by their comrades in Soviet Russia. The people whose departure was in question were their friends and political partners. But their concern went beyond the personal. Many of the Soviet Zionists who had arrived in Palestine continued to see themselves as members of their old movements, and they sought to secure for them a role among the parties of Labour Zionism in Palestine and in Europe.[2] Following a tradition developed by the oppositional parties in Russia, each of the Soviet Zionist movements established a delegation abroad to represent its interests in the combative political arena of Labour Zionism.[3] And they naturally sought to increase their numbers in Palestine by arranging for the immigration of their members. Once earlier expectations for massive immigration from Soviet Russia were crushed, the movements turned their attention to the singular opportunity opened by the substitution emigration.

The delegations abroad of Hitachdut (STP), Zionist–Socialist Party (TzS), and Ha-Shomer ha-Tza'ir played a crucial role in the early stages of the substitution immigration and continued to act as its impassioned advocates. In 1924 and 1925, before Pompolit had come to play a central role in the communications between Moscow and Palestine, the movements and their delegations were the most effective channel for names of those arrested and granted substitution.[4] In February 1925 they were given a de facto monopoly on nominations for substitution immigration, when the Immigration Centre announced that, given the avalanche of individual requests, only lists submitted by the movements would be accepted.[5] In those early years, the movements also coordinated the delivery of immigration permits to contacts in Moscow, often by telegraph.[6] And they were frequently called upon to provide the Immigration Centre with information to be used in putting forth requests to the PZE and the Palestine Government.[7]

The delegations also acted as a pressure group within the Histadrut and, through it, demanded financial and diplomatic help from the Zionist Executive. At first, the socialist Zionists from Soviet Russia, like many in the socialist party Achdut ha-Avoda with which they affiliated themselves in Palestine, were hesitant to draw attention to Soviet repression.[8] Ben-Gurion had warned in 1923 that open criticism would jeopardize the measure of legality attained by Hechalutz, and he called on labour Zionism to work through the Soviet government, not against it.[9] But things began to change after the mass arrests of September 1924 and more so in early 1925, when another wave of arrests indicated a Soviet decision to liquidate the Zionist movements.[10] Thus, in spring 1925, the conference of Achdut ha-Avoda resolved to launch a campaign to help the Zionists in Soviet Russia.[11] There followed a week of lectures, assemblies and fund raising, and a special issue of the party's weekly, *Kuntres*, dedicated 'to the work of helping our comrades in Russia'.[12] Thereafter, the socialist movements worked largely through the Histadrut and its parties and used their publications to broadcast the struggles and sufferings of their comrades in Soviet Russia.[13] The Histadrut's leaders seemed willing to harness its moral authority for the protest against Soviet harassment of Zionists.[14] This mobilization reached a peak after the liquidation of Legal Hechalutz in 1928, with editorials in *Davar*, mass meetings, speeches by the leaders of the Histadrut and resolutions by various parties and organizations.[15]

A special role was reserved for *Davar*, the Histadrut's daily paper established in 1925. During its first few years, the paper carried regular reports about the legal and political conditions for Zionist work in Soviet Russia and long debates about the veracity of Soviet pronouncements of tolerance towards Zionism.[16] Arrests of Zionists were reported on its front page in bold headlines.[17] And when the number of exiled Zionists increased, the paper published many articles about the harsh realities of their lives, the obstacles put by the Soviet authorities in the way of their 'substitution', and the need for immediate action by the Histadrut and the Zionist Organization.[18] The pages of *Davar* were also used to broadcast public protests against Zionist persecution in Soviet Russia and advertize the organizations dedicated to help those arrested.[19]

In 1927, to maximize their effectiveness, the delegations abroad of all the Soviet Zionist movements established a joint organization, the United Committee to Help the Prisoners of Zion (Ha-Va'ad ha-Me'uchad le-Ezrat Shvuyei Tziyon).[20] Together, the individual movements and the committee lobbied the delegates to the bi-annual Zionist congresses with reports on the worsening conditions for Zionist work in Soviet Russia and demanded moral and financial support.[21] At least some of these reports were forwarded to the Palestine Government and to contacts in the British government.

Unlike the socialist and labour movements with their delegations abroad, former activists of the General Zionist Organization lacked a solid organizational base in Palestine and a platform for publicizing their concerns. It took until 1929 for them to launch an initiative of their own – a meeting in Tel Aviv, attended by the old lions of Russian Zionism and Hebrew culture, who founded Magen, the Society to Help Those Persecuted for Jewishness, Zionism and All National Matters in Soviet Russia.[22] Later joined by members of other movements, Magen issued bulletins, organized public meetings and lobbied with the PZE on behalf of immigrants from Soviet Russia. Its second meeting in June 1931 called on the Zionist Congress to provide urgently needed funds.

The General Zionists and the labour Zionist parties differed over goals and rhetoric. The General Zionist position was anti-Soviet from the outset and their rhetoric was based on the moral repugnancy of the regime and, conversely, the moral right of its victims. Their practical demands were for aid in emigration and, occasionally, for those languishing in Soviet prisons or exile. A memorandum sent to Chaim Weizmann in late 1924 by Se'adia Goldberg, a General Zionist activist who came as a substitution immigrant in July 1924, was emblematic in anchoring the claim of Russian Zionists in their outstanding contribution to the settlement in Palestine and their persecution by the Soviet regime.[23] 'If we do not secure permits to enter Palestine, we will bear the blame for the lives of exiled comrades. The honour of the Zionist Organization is in question here. Eretz-Yisrael can not close its doors to Zionists who are giving their souls for it'.[24] Besides, he added, Russian Jewry, 'hitherto, the heart of world Jewry', could still offer the Zionist enterprise excellent 'human and national material'. First, among these potential builders of the Zionist dream he counted the imprisoned and exiled Zionists. Two years later, Israel Rozoff, another General Zionist, urged the London Executive to create a secret fund of £2,000 to facilitate the emigration of those who were allowed to leave and help those who are condemned to 'hunger and torture in exile'. Only in this way could the Zionist Organization change the harmful perception that world Zionism had deserted its brothers in Soviet Russia.[25]

The labour movements sought clearer political goals. Each asked for funds to support the work of comrades still active in the underground in Russia and those in exile. In May 1926, for example, Yitzchak Vilenchuk, representing STP, presented to the Zionist Executive the following demands: £2,000 for the immigration of 50 Zionists in Tobol'sk; £8,400 to provide 700 imprisoned Zionists with £2 monthly help; £3,000 to enable 1,200 *chalutzim* to settle as shareholders in colonies in Crimea and £400 for a printing plant for STP to print its journal and other propaganda.[26] A year later, Nachum Verlinsky, writing for both STP and Illegal Hechalutz, asked the London Executive for a monthly allocation of £50 to support underground activity in Soviet Russia, including a network for smuggling Zionist literature and spreading propaganda among Jews settling on the land in Crimea and the work of two Hechalutz activists who had returned illegally from Palestine to help the movement in its difficult hour.[27] As we shall see, the Zionist Organization was more sympathetic to the minimalist requests of the General Zionists than to the activism of the labour and socialist movements.

The historic schism between socialists and toilers[28] also entered this discourse and at times affected the treatment of those fleeing Soviet repression. A multilateral correspondence in autumn 1924 and early 1925 reveals an example of discrimination by the Latvian Zionist Organization and the Riga Palestine Office against members of Legal Hechalutz who had taken great risks in crossing the border illegally into Latvia. They were allowed to languish in government quarantine and were otherwise denied help on their way to Palestine. When World Hechalutz in Berlin and the Zionist Executive in London inquired about the case, the Riga Zionists responded by casting doubt on Legal Hechalutz.

> The question arises, is such an element desirable in Palestine? Are they not imbued with the doctrine of Zinov'ev to such an extent as to leave no love for Palestine itself? They may perhaps do more harm than good in that country, [though] it must be admitted that some of them speak and write Hebrew very well.[29]

The socialist wing of Soviet Zionism answered in kind, blaming Illegal Hechalutz and the STP for the Latvians' conduct.[30] Typically for those years, the conflict was settled in favour of Legal Hechalutz: representatives of the Histadrut persuaded the PZE to reject any distinction between 'red' and 'white' *chalutzim*.[31]

Notwithstanding these divisions and conflicts, the Zionist Soviet movements succeeded in raising sympathy for their comrades in Soviet Russia, especially among the labour movement in Palestine. But in the 1920s and 1930s, the case of the imprisoned Zionists had to compete for the attention of the Zionist Organization with the troubled news from Jews all over Eastern Europe. When one Zionist Congress after another failed to take decisive action and commit major funds, the Soviet Zionist movements and the Histadrut at large saw this as betrayal. 'What name can we give it?' asked an article in the moderate Labour journal *Ha-Po'el ha-Tza'ir* in February 1929. 'How can one explain such criminal apathy?'[32]

Beyond its practical impact on the substitution immigration, the campaign on behalf of the Zionist prisoners left an imprint on Zionist collective memory. To attract attention and sympathy, some advocates went beyond the facts of the prisoners' plight, publishing exaggerated claims about their numbers ('thousands') and unfounded assertions regarding a Soviet policy of 'physical destruction'. The pleas for help were frequently suffused with high pathos, as in the following passage from *Davar*:

> Thou shall visit the memory of your dear brethren in Russia and behold the hundreds and thousands of youth who languish in torment when their hands are unsoiled by any transgressions. Their desire for salvation is hidden in the crannies of prisons, in dampness and cold, in detention camps, in the steppe of Kirgiziia and on frost-covered islands. Poverty, scurvy, consumption and a weak heart do not deter them. Torn from their homeland, in loneliness, they nurture with their own warmth the vision of a [national] renaissance.[33]

Over the years, this campaign established the image of the Soviet Zionists as 'Prisoners of Zion' – entombed in the arctic penal camps of Stalin's era for their commitment to Zionism – an image that would gain resonance especially after the birth of the Jewish state.

The Immigration Centre (Merkaz ha-Aliya)

The Immigration Centre was the next link in the organizational chain that facilitated the immigration of Zionist convicts. Set up in 1919 as an *ad hoc* response to the spontaneous wave of pioneers rushing to Palestine after the establishment of the British Mandate, the Immigration Centre became part of the Histadrut's fast growing institutional framework.[34] From the beginning and for years to come, it provided badly needed services to those arriving at the shores of Palestine with little practical experience or knowledge of the country. It directed pioneers to where work was available, lobbied with the PZE and the British authorities for work in the development projects undertaken during the first years of British rule, helped organize individual immigrants into labour detachments and pressured the PZE for land (purchased from its Arab owners) on which to settle collective groups.[35] At many junctures, it pushed the reluctant leadership of the Zionist Organization to speed up immigration, sending representatives in 1919 to work from within the 'Palestine Offices', which the Zionist Organization had begun to establish in Vienna, Warsaw, Kishinev and other Jewish centres (excluding Soviet Russia). Later, it worked with the central office of Hechalutz in Berlin to ensure a flow of working immigrants for the settlement in Palestine, especially those whose training and commitment guaranteed their willingness to 'put themselves at the disposal of the Histadrut'.[36]

When substitution immigration began in 1924 and 1925, the Immigration Centre quickly came to occupy a central role in its organization. Numerous requests flowed into its office in Tel Aviv from friends and relatives in Soviet Russia and in Palestine, announcing the arrest or sentencing of loved ones. The delegations abroad provided more orderly lists of their movements' arrested members, collected through their contacts in Soviet Russia and from members arriving in Palestine.[37] After 1926, news from Russia usually came through Pompolit. Although we have no direct knowledge of how this contact was established, it was most likely through Chaim Halperin, an agronomist and long-time member of Legal Hechalutz who had spent time in exile before receiving substitution. In Palestine, where he arrived in 1925, Halperin headed the Secretariat of the Immigration Centre and personally carried on the correspondence with Peshkova and Vinaver. Through his contacts, the Immigration Centre became, in effect, the link between Pompolit, the PZE and the British government, or as he described it, 'the de facto Palestine Office for Russia'.[38]

That the Immigration Centre occupied so central a place in arranging the substitution immigration (and immigration to Palestine in general) was due to two political realities of the Jewish settlement in Palestine in the 1920s and 1930s: the ever increasing strength and influence of the labour parties and the Histadrut, and the predominance of parties and individuals with roots in the Russian Empire and its Soviet successor. The Immigration Centre itself was led by a succession of recent immigrants who had done their political apprenticeship in the left wing of Soviet Zionism – Yisrael Mereminsky (Marom), Levi Dobkin and Chaim Halperin, all members of Legal Hechalutz.[39] The personal and political orientation of these men, like the goals of the Histadrut at large, assigned priority to the immigration of committed Zionists from Soviet Russia, especially those sentenced to exile.

The importance the leaders of the Immigration Centre placed on the substitution immigration can be gleaned from a letter sent on 29 October 1924 to World Hechalutz in Berlin. Reporting on its first meeting with the PZE's Departments of Immigration and Labour, the Immigration Centre announced proudly the release by the British Controller of Immigration of 61 certificates specifically for Zionists imprisoned in Soviet Russia. Never mind that an equal number of certificates would have to be subtracted from the usual six-month allocation.[40]

> We think that all of those coming from Russia, having been arrested and released, will be true *chalutzim* and will accept any and all work. These are the kind of people we have been waiting for and we must use this opportunity [to bring immigrants] from Russia, although we know how valuable it is to increase immigration from all of Eastern and Central Europe.

From the vantage point of Hechalutz in other countries (even in Soviet Russia) the preference given to the substitution immigrants remained problematic. M. Bogdanovsky, writing for the Central Bureau of World Hechalutz, pleaded for urgent action on behalf of all Russian *chalutzim*:

> In no way can we be satisfied with certificates only for those arrested. There is a pressing need for immediate *aliya* of 200–300.... From our comrades' letters we know that hundreds of people are flocking to Moscow from all over Russia and are waiting for... *aliya*. Every day, telegrams and letters from every corner of the country reach the Russian Centre of Hechalutz. About 300 *chalutzim* have left their work and stand mobilized. The rush for *aliya*

grows as interest in Eretz Yisrael increases, and unless you are able to provide 300 certificates for free immigration, Hechalutz and all of the youth in Russia are condemned to suffocate. We know the difficulties, but don't allow yourselves to be silenced until you have saved the movement in Russia.[41]

Perhaps in response to these exhortations, the Immigration Centre pressed the PZE to demand from the government not to include the substitution certificates in the regular quota, 'Otherwise, as you know, there will be no way of sending certificates for *chalutzim* not arrested'.[42] But substitution immigration continued to be given highest priority.

To achieve its aims, the Immigration Centre sought to oversee the flow of information and requests for substitution immigration. In autumn 1924, it insisted and repeatedly reminded Hechalutz in Russia and the movement's world centre in Berlin that all such requests had to go through the Immigration Centre. There were practical justifications for this arrangement, since on several occasions individuals and groups had forwarded incorrect information to the PZE, claiming the status of prisoners for people who had not been arrested at all, and these mistakes had caused much displeasure among British immigration officials.[43] But the Immigration Centre was also seeking to secure its political grip, as indicated in a letter signed by Mereminsky. The Immigration Centre, he insisted, would '(1) not pay attention to telegrams and letters regarding immigration from non-workers; (2) not handle any such matters, when the request is addressed both to us and to the Zionist Executive'.[44]

Within the Immigration Centre, there was political friction as well, most notably between its activists who had belonged to TzS and Legal Hechalutz on the one hand, and Binyamin Vest, representing Hitachdut and Illegal Hechalutz on the other. Vest accused the Immigration Centre of 'indifference' to the vital needs of Illegal Hechalutz for Immigration Certificates, especially for those in prison, and established his own direct contacts with the Zionist Executive in Jerusalem. The Immigration Centre attempted to reassure Vest that it was committed to all of the *chalutzim* coming from Russia 'without distinction to which organization they belong' and, at the same time, instructed him to cease his separate activities, thus provoking his resignation from the Centre in early 1925.[45] Accusations of bias by the Immigration Centre came from other leaders of the non-socialist wing of Soviet Zionism as well, accompanied at times by claims that their movements suffered disproportionately at the hands of the Soviets.[46]

In truth, the Immigration Centre and the Histadrut's Executive Committee had shown some bias towards the socialist wing of Hechalutz. Ben-Gurion expressed his sympathy in public and in private letters.[47] A letter from Mereminsky to Hechalutz in Berlin reported that the Immigration Centre had held extensive discussions with the newly arrived Levi Dobkin about the ambitious plans he and his comrades in the Central Committee of Legal Hechalutz in Moscow had developed for close economic ties between the labour movement in Palestine and the

Soviet Union. Details of these plans – anathema to the non-socialist wing of Soviet Zionism – were distributed to all Histadrut agencies with a request to 'discuss these [plans] and make practical suggestions for economic and commercial ties between the Histadrut and Soviet Russia'.[48]

The friction with regard to immigration priorities should not be surprising. Not only was there deep hostility between the socialist and toilers' wings of Soviet Zionism, but the political arena they found in Palestine was rife with ideological differences and political manoeuvres. In this environment, the socialist, collectivist wing of Soviet Zionism struck alliances with the like-minded, powerful wing of the labour movement and used the positions it thus gained to fight for immigration of socialist Zionists and substitution immigration in particular.[49]

Significantly more damaging than such squabbling was the inability of the Immigration Centre to obtain appropriate funding. Besieged by calls for help, the Centre made numerous appeals to the Zionist Organization and its Palestine Executive, repeatedly arguing that it was not enough to allocate Immigration Certificates for Russian *chalutzim*.

> Those who have given up and left the ranks of Hechalutz complain about one thing only – no hope of *aliya*. There are hundreds of members with 7 or 8 years of membership and work in Hechalutz, whose turn for *aliya* has not yet come, not for lack of certificates, but because they have no means to pay for their way.

Of two hundred certificates sent to Russia in June 1926, only a handful had been used by late November.[50] Even when matters came to those in prison and exile, complained Chaim Halperin in May 1928, 'the negotiations we have been carrying out with the PZE for four years have not produced positive results'.[51] When funds appeared, there were thorny questions of distribution and accountability. Thus, for example, upon the allocation of 900 Egyptian Liras in 1925, the Financial Department of the PZE demanded 150 separate receipts from Zionists in Soviet Russia, which led to lengthy, and at times acerbic exchanges between the Immigration Centre, the Immigration Department and Hechalutz in Russia.[52]

The more impatient advocates of the Zionist exiles blamed the Histadrut itself for the failure to raise funds. The Executive Committee had found a way to send money to help the striking miners in England, wrote one of them, yet failed to do the same for the prisoners and exiles in Soviet Russia.[53] But the Immigration Centre and the delegations of the movements continued to see the PZE as responsible, not only for the lack of funds but also for the insufficient allocations of certificates for political exiles from Soviet Russia and its lack of 'energetic intervention' to ensure accommodating British policies.[54] Did the Zionist Executives deserve the accusations levelled at them? To answer this question, the following pages will examine their actions in London and Palestine, in the three areas crucial to immigration from Soviet Russia – funding, negotiations with the British and allocation of certificates.

The World Zionist Organization and its executives

Whereas personal and political loyalties disposed the representatives of the Russian Zionist movements and the Immigration Centre to give priority to the immigration of Russian *chalutzim*, especially those arrested and exiled for their Zionist work, matters were more complicated for the World Zionist Organization and its London-based Executive. The Zionist Organization weighed the cause of Soviet Zionists against the needs of Zionists elsewhere, especially in Eastern Europe. The issues were often budgetary, as the Zionist Organization struggled to balance its meagre funds and was repeatedly called upon to help the settlers in Palestine through prolonged economic crises and periods of unemployment.[55] Indeed, much of the attention of the Zionist Organization during this time was absorbed in the effort to enlist immediate help from wealthier Jews in Western Europe and the United States, and to establish the Jewish Agency, through which the world Jewish community was expected to support the settlement in Palestine. In addition, the Zionist leadership, especially Chaim Weizmann, was engaged in a fight against the many voices in Britain demanding to limit or even reverse the Balfour Declaration's promise of a 'national home', and this required, among other things, that the substitution immigration not be viewed as a channel for slipping Communist agents into Palestine.

How, then, did the World Zionist Organization respond when confronted with the immigration of Zionist prisoners? Here we need to separate Weizmann's response (after all, his own roots were in Russian Zionism) from that of the organization at large, and to distinguish between sympathy and effective action. In spring 1924, the Central Office of the Zionist Organization in London began to receive urgent calls for diplomatic and financial help to save Soviet Zionists from exile – from Berlin, Moscow, Kovno, Palestine and from as far away as South Africa. Its response came quickly and readily. By the end of June 1924, not less than five meetings of the London Executive had discussed the issue of substitution immigration, treating the matter as urgent and extraordinary. Initially chaired by Weizmann, the meetings resolved to undertake several diplomatic actions, including a planned meeting with Deputy Commissar for Foreign Affairs Maksim Litvinov and contacts with the newly appointed British Minister to Moscow. Typical of what would ensue, the financial issue was harder to settle, though at Weizmann's behest, funds were committed to help the Soviet Zionists – in the hope of collecting contributions from sympathetic individuals.[56]

Over the next decade, the same pattern repeated itself. The Zionist Organization continued to meet occasional *ad hoc* requests on behalf of the Zionists in Soviet Russia, exempting them from the general rule of self-funding by the national federations of the organization. In the first half of 1926, for example, the Executive approved several *ad hoc* requests for funds, among them £250 to enable thirty-three convicted Zionists stranded in Odessa to reach Palestine, and £120 in reimbursement for the Joint Distribution Committee (JDC) to cover its aid to Zionists in Moscow.[57] But the Zionist Executive rejected demands from the Histadrut and the Soviet Zionist movements for funds to continue Zionist work in Russia or to provide steady funding for the immigration of Russian *chalutzim*.[58] This seeming lack of support must be viewed in the context of the Zionist Organization's perennial budgetary difficulties. Its central fund raising efforts fell far short of their ambitious goals, thus disappointing the earlier hopes of the Zionists themselves – and the British – for a rapid development of Palestine with the help of Western Jewish money and East European immigrants. Local Zionist federations were left to raise their own funds for all activities, including the emigration of *chalutzim* to Palestine. Indeed, the modest help granted by the World Zionist Organization to the Soviet Zionists was extraordinary and could only be explained in the absence of a Russian Zionist federation.

Several efforts were made to raise special funds for the Soviet Zionists. In 1926, Weizmann proposed that a sub-committee be set up to examine ways to meet this need and offered a first donation of £50 from his own pocket towards the effort.[59] The fifteenth Zionist Congress (1927), created a 'Central Committee to Aid Russian Zionists', headed by Leo Motzkin, but it proved ineffectual, leading Chaim Halperin and other veterans of the Soviet Zionist movements to turn for help to the non-Zionist Jewish organizations – the JDC, the Jewish Colonization Association (JCA)[60], and the Jewish Emigration Society[61] – to the annoyance and embarrassment of the Zionist leadership.[62] In another effort, in 1929, the Zionist Organization's Central Office in London issued an appeal to all members of the Executive Committee and to the federations and fractions to make contributions to a special secret fund of £2,000, exhorting them to fulfil their 'moral obligation' and prove the solidarity of the Zionist movement.[63] Some funds were collected and used to help the exiles and immigrants, but they were woefully inadequate. In 1933, for example, the Jewish Agency's Executive admitted that immigration from Russia was severely hampered by the shortage of funds to cover passport fees and transportation.

> As a result, even many of the exiled Zionists who had been released could not use the certificates already approved by the Immigration Department and remained in their Siberian exile for several additional years, in a state of severe suffering and despair of any help.[64]

It is possible that donors were fearful that paying the high fees demanded by the Soviet authorities may inspire other governments to exploit the plight of their Jews for revenues. But the greater impediment to raising funds for Soviet Zionists lay in the differences over priorities and goals that separated the Zionists from the Jewish communities with the financial means to help. For the non-Zionist leaders of American Jewry, the paramount task was to help the Jewish population in Eastern Europe and Soviet Russia, and they disagreed strongly with the Zionists' negative assessment of the project of Jewish agricultural settlement in Crimea. In 1925 and 1926, the relatively small American Zionist movement was entangled in an acrimonious conflict with the non-Zionist leadership over the plans of the JDC to raise $6.5 million for the Crimean Project and other efforts to draw Jews into productive labour in Soviet Russia while providing a much smaller sum for agricultural and industrial development in Palestine.[65] Weizmann, eager to gain American financial support, sought to minimize these disagreements by presenting to the American leaders a more positive view of the so-called Crimea Project (while painting it rather darkly in other communications).[66] His public position on this issue cost him the support of many in the World Zionist Organization, yet there can be no doubt about his deep commitment to the cause of Russian Zionism. That commitment as well as his dependence on the American Jewish public led him to urge Chaim Arlosoroff, head of the Zionist organization's Political Department, to take up the matter during a trip to the US in fall 1928:

> The struggle which our young comrades are carrying on for years in the face of the most adverse and dangerous conditions is indeed one of the most encouraging signs of the strength and vitality of Zionism among the Russian Jewish youth. I consider it a sacred duty of every Zionist to lend his aid to the young Zionist Federation [Hitachdut/STP] which has grown up in Russia under exceptionally hard and unfavourable conditions.
>
> I trust you will succeed in awakening the interest of American Jewry in this important problem and that your appeal for financial support will meet with the response it deserves.[67]

Writing to the Jewish American philanthropist Felix M. Warburg in May 1930 to defend the special collection for Soviet Zionists, Weizmann explained the need to help 'numerous Zionists who are pining away in imprisonment' and stated emphatically that 'quite apart from all political considerations, [it is] simply a duty of fraternity and solidarity to go to their aid'.[68] To his disappointment, the campaign to raise funds for Soviet Zionists, and indeed the very work of the Zionist underground in Soviet Russia, encountered criticism by at least some American Jewish leaders, who feared it might endanger not only the Zionists themselves, but also the work of the non-Zionist organizations, especially the JDC.[69]

Money was not the only pressing need. Emigration from Soviet Russia to Palestine required British help in providing Immigration Certificates and easing immigration regulations to accommodate the rigidities and arbitrariness of Soviet procedures. In addition, the Soviet Zionists repeatedly demanded from the Zionist Executive to secure the posting of *aliya* emissaries from Palestine in Soviet Russia. Direct negotiations on this issue between the London Executive and the Colonial Office took place in September 1927, in the wake of severance of Britain's diplomatic relations with the Soviet Union.[70] Here, too, Weizmann's dedication to the Soviet Zionists was strong. In December 1929, amidst the unfolding Arab disturbances in Palestine, he found time to meet with Esmond Evory, the newly appointed British ambassador to the Soviet Union and submitted to him a memorandum on Soviet persecution of Zionism. He also met with Labour's Foreign Secretary, Arthur Henderson, and remained in touch with his Parliamentary Under-Secretary of State, Philip Noel-Baker.[71] For reasons discussed later in this chapter, Weizmann's efforts *vis-à-vis* the British were less successful in these final years of the substitution immigration. Still, his greater failure, and that of world Zionism, was the inability to raise funds that could have allowed more Zionist prisoners to leave for Palestine when it was still possible.

We now turn our attention to the Zionist Executive in Palestine (PZE) which had the principal responsibility for representing the Jewish community in Palestine on all matters of immigration. It negotiated with the Palestine Government over the number of Immigration Certificates allocated by the government every 6 months under the Labour Schedule,[72] distributed the certificates among the Palestine Offices located in countries with significant Jewish population and guaranteed subsistence for all immigrants in the category of 'labourers' during their first year in the country. While acting as an agency of the Zionist Organization (until the creation of the Jewish Agency in 1929), the PZE was influenced by its immediate environment in British Mandatory Palestine and cooperated closely, in spite of often being at odds, with both the Palestine Government and the Histadrut. Like the Immigration Centre, it considered the Russian immigrants to be not only prisoners in the cause of Zionism, but also highly qualified for life in Palestine and, in particular, for the manual labour that was often the only source of livelihood available to the labourers nominated by the PZE.[73] At the same time, the PZE seems to have shared the wariness of the British about possible infiltration of Soviet agents among the immigrants and devised ways to ensure against it.

From the middle of 1924 and for as long as Soviet Zionists were able to obtain permission to leave Russia, the PZE and its Immigration Department worked hard to ease the path of their immigration, relying for the most part on the Histadrut's Immigration Centre to supply names and other required information (age, address and political affiliation) for the prospective immigrants. While the PZE's Political Department negotiated special dispensations for Zionist refugees from Soviet Russia with the British Government in Palestine, the task of the Immigration Department was to channel information from the Immigration Centre to the Palestine Government's Immigration Department (later Section), provide the assurances and guarantees required by the government and press for a speedy delivery of the Immigration Certificates to the individuals named.[74]

One measure of the PZE's concern for the Russian Zionists in general, and those arrested in particular, is to be found in its allocation of Immigration Certificates. For most of the decade of substitution immigration, the number of certificates allowed by the British fell short of the demand from the *chalutzim* waiting in Eastern Europe and from Palestine residents wishing to bring over their relatives. For that reason, the PZE's distribution of the certificates among the various countries was a sensitive issue, especially during years of restricted immigration. Correspondence between the PZE's Immigration Department and other Zionist offices shows the percentage of certificates allocated to Russia to have been maintained at a high level for a number of years, standing at around 25 per cent during the boom years of 1924–1925.[75] This policy was confirmed at a meeting between the head of the Immigration Department and all the Palestine Offices in Europe (17–25 March 1925) where 2,000 out of 8,000 certificates made available by the government were allocated to Russia and an appeal was made for special budgetary attention to the 'extraordinary conditions of immigration' from there.[76] A special section on immigration from Russia, included in the Immigration Department's annual report to the PZE (covering the period May 1925 to May 1926), stressed the 'urgent need' to post a representative to and increase immigration from Russia, 'especially of those who are persecuted for Zionist activity'.[77] In early 1927, a year of reduced immigration, one-fifth of the 500 available certificates was earmarked by the Immigration Department for Zionists arrested in Soviet Russia. Figures for 1931 likewise hovered around 20 per cent.[78]

The Immigration Department played an important role in some of these decisions. For example, only pressure from the Department's head could explain the readiness of the Palestine Offices to support the large allocation of certificates to Soviet Zionists when the pressure for emigration ran high in their own countries. The Immigration Department was also responsible for the demand for special funds, because the only alternative would have been for the funds to come from its own budget. However, the policy of favouring the Soviet Zionists in the distribution of certificates came from the highest levels of the Zionist organization in Palestine – from its Executive Committee and the PZE's strongest man, Colonel Frederick H. Kisch.[79]

Serving as Chairman of the PZE's Political Department, Kisch was a most surprising advocate and an effective facilitator of the substitution immigration. A scion of an established Anglo-Jewish family and the son of a British official in India, a decorated field commander and a budding diplomat, he had no prior commitment to Zionism. Nevertheless, he was asked by Weizmann to come to Palestine to spearhead the Zionist political effort under the Mandatory government. His selection was due, no doubt, to his intimate familiarity with the British colonial service and easy rapport with high ranking British officials in Palestine. During his term (which ended in 1931), much of the communications with the British over issues of immigration was conducted in Palestine itself and only on rare occasion was the London Executive asked to intervene with the Colonial or Foreign Offices. At least for the duration of Herbert Samuel's years as High Commissioner for Palestine (1921–1925), Kisch met with him weekly.

Kisch was introduced to the issue of substitution immigration from Soviet Russia by David Ben-Gurion, who presented him with the first requests for entry permits for Zionist convicts on 9 and 26 June 1924. Ben-Gurion also discussed with Kisch the Soviet government's reported decision to allow the immigration of 500 *chalutzim*, which he believed would 'open the door for fresh immigration of the best type'.[80] This must have been welcome news to Kisch, who had approached the British Government of Palestine a few months earlier to request that pressure be put on the Soviet government to allow Jewish emigration to Palestine, or, at the very least, permit the departure of 500 members of Hechalutz.[81] Kisch followed closely the arrival of the *Novorossiisk* (the first boat carrying substitution immigrants) on 14 July 1924 and immediately met with one of its passengers, the General Zionist activist, Se'adia Goldberg.[82] He went on to inform Weizmann about Ben-Gurion's upcoming visit to London to discuss large-scale immigration of Hechalutz members from Russia, stating clearly his own position: 'I regard this as a most important matter, and hope you will give it all your support'.[83] Kisch made immigration from Russia the central issue of his report to members of the Executive (dated 13 August 1924), with special attention to the two issues that had occupied him throughout the spring and summer of 1924 – immigration of 'victims of persecution' and the reported Soviet permission for 500 *chalutzim* to leave for Palestine.[84] During 1925, Kisch tried in vain to convince the British to post a consul in Odessa. He then pressed Weizmann in London to secure the establishment of a Palestine Immigration Office at the British Legation in Moscow.[85]

Kisch did not hesitate to intervene personally with the British authorities on smaller matters as well. The PZE files hold numerous examples of requests from officials of the Immigration Department and the Secretariat for his intercession with British officials to speed up the process. The first step usually involved the Controller of Immigration,[86] but when necessary Kisch pursued matters with the Chief Secretary or the High Commissioner and even the Colonial Office in London.[87] Most frequently, the requests were for additional certificates for political refugees, preferably outside the regular Labour Schedule, or to speed up the Government's approval process. Kisch also intervened on behalf of requests to extend the validity of the certificates (the time allowed for the issuing of visas and for entering Palestine) and to arrange a convenient place for the immigrants to receive their certificates. In general, Kisch credited the Palestine Government with providing 'generous facilities in regard to making possible the immigration to Palestine of political refugees'.[88] Indeed, as we shall presently see, the British had their own reasons to show sympathy for the imprisoned Zionists, though their actual policies reflected several different considerations. Throughout Kisch's years at the PZE, his interventions helped shape the British approach to the immigration of Soviet Zionists, even as his exchanges with officials of the Palestine Government became increasingly strained.

The British Government in London and in Palestine

For the British Mandatory Government of Palestine, there was hardly an issue as politically volatile as immigration. Central to the promise of the Balfour Declaration to build a 'national home' for the Jews in Palestine, it was a frequent flash point in the conflict between the Zionist settlers and the Arab residents of Palestine. When violence between Arabs and Jews broke out in May 1921, it brought to an end a year of immigration free of British controls. Thereafter, political considerations played a significant role in British regulation of immigration into Palestine. But the record of immigration during that first year demonstrated that economic issues were just as crucial and that the Zionist Organization itself played a role in restricting immigration: of the 70,000 immigrants agreed on by the British and the Zionists for that year, only 8,000 or so actually came, a disappointing result that was largely due to the inability of the Zionist Organization to raise the funds needed to provide work and housing for the new arrivals. Throughout the Mandatory period, economic and political considerations continued to dictate British policy on immigration in general. With regard to the substitution immigration, the policy and regulations were also determined by the changing tenor of Anglo-Soviet relations.

To a surprising degree, the British stand on the immigration of Zionist convicts from Soviet Russia was amenable to pressures and pleas from the Zionist leadership. This responsiveness, especially during the first year of substitution immigration, gave rise to rumours about an agreement, reportedly concluded between the British and Soviet governments, to allow massive immigration to Palestine, including an initial group of 500 *chalutzim*.[89]

An examination of Zionist and British documents puts that myth to rest, exposing it as an embarrassing misunderstanding. Its origins can be traced to a message sent by the British Mission in Moscow (21 May 1924) announcing Soviet 'sanction' for the departure to Palestine of 500 young men, ages 20–28, under the auspices of Hechalutz, with additional groups to follow.[90] It was later determined that British Mission personnel had obtained this information not from Soviet government officials, but from Professor Tzvi Belkovskii, a leading General Zionist who was soon arrested and allowed to depart for Palestine. In the meantime, news of the Soviet 'sanction' had been passed by the Chief Secretary of the Palestine Government to Kisch, and arrangements were swiftly made for the Chairman of the Histadrut, Ben-Gurion, and for Levi Shkolnik (Eshkol) to travel to Moscow to arrange the selection of suitable immigrants and coordinate their transportation to Palestine.[91] Ben-Gurion made this his highest priority, hoping to bolster Hechalutz with prospects for emigration, and believing that Russia could offer Palestine not only the best human resources, but valuable economic cooperation.[92] In fact, it was Ben-Gurion and Shkolnik's request for Soviet visas that exposed the misunderstanding. Before arranging for their visas, the Soviet Trade Delegation in London asked for further details about the alleged sanction for mass immigration. The request was passed by the Zionist Executive in London to the Colonial Office, from there to the Foreign Office, and finally to the British Mission in Moscow, which responded on 5 March 1925, explaining that the rumour had originated in a conversation with Professor Belkovskii.[93]

From the British side, notwithstanding the bureaucratic confusion, there was strikingly close cooperation with Zionists at all levels, both in Jerusalem and in London. Indeed, British support for substitution immigration was the rule in the early years (1924 through 1926). Decisions were usually made by the Palestine Government, with little interference from the Colonial Office. Exceptions to standing immigration regulations were readily granted by the Controller of Immigration, Albert Hyamson, upon request from the PZE.[94] The Palestine authorities were in direct communication with the British Chargé d'Affaires in Moscow, thus facilitating – and controlling – the distribution of visas for Palestine. Permits to enter Palestine under the Labour Schedule were frequently sent by telegraph to Moscow. From mid-1924 and throughout 1925, Hyamson and the Chief Secretary also joined the effort to convince the Foreign Office to appoint a British Consul to Odessa, 'to serve the largest concentration of Jews in Ukraine and southern Russia'.[95] Several explanations account for this relatively congenial state of affairs: the general conditions for immigration into Palestine at that time, the pro-Zionist stance of the Palestine Government, and Britain's brief thaw with the Soviet Union, as seen in the establishment of diplomatic relations in January 1924 and the posting of consuls to Moscow and Leningrad. Although Anglo-Soviet relations did not rise at that time to the ambassadorial level – indeed they began to sour by autumn 1924 – the presence of British consuls in Russia greatly eased the establishment of substitution immigration during its earliest stages. British officials in Palestine and in London supported the continuation of this arrangement even when relations began to decline, motivated both by the wish to encourage and support critics of the Soviet regime and, in some cases, by true sympathy for suffering Zionists.

British immigration policy was tied to economic and employment conditions from the issuing of the White Paper of 1922 and the immigration regulations that followed. These divided prospective immigrants into several categories (i.e. tourists, religious students, persons of means, members of professions, dependents of Palestine residents and unskilled labourers), with the PZE retaining significant responsibility only for labourers (Category E), whose livelihood it was required to guarantee for a period of one year.[96] Twice annually, the government announced the Labour Schedule for the following six months (April to September and October to March). The PZE was invited to present estimates regarding the number of labourers needed, but the final decision was reserved for the Controller of Immigration, often in concert with the High Commissioner for Palestine or his Chief Secretary. An unexpected economic upswing from spring 1924 through autumn 1925, followed by massive immigration of families with at least modest capital from Poland beginning in 1925, created unprecedented employment opportunities (in construction and in the developing tobacco growing industry) and produced more flexible Labour Schedules.[97]

Beyond economic conditions, the government's willingness to undertake special measures to facilitate the substitution immigration was rooted in the personal convictions and sympathies of its high-ranking officials. In 1924 and the first half of 1925, the High Commissioner for Palestine was Herbert Samuel, his Secretary (later Assistant Secretary in the government's Secretariat) Max Nurock, the Attorney General for the government Norman Bentwich and the Controller of Immigration Albert Hyamson – all Anglo-Jews whose selection for these positions owed much to their declared Zionism.[98] In time, Kisch would come to consider Hyamson an opponent who was bent on discrediting the PZE, but in 1924 he found both Nurock and Hyamson exceedingly willing to cooperate.[99] A letter from Hyamson to the Moscow Chargé d'Affaires from that time is exemplary: though suspicious that the Zionists' claims of persecution were deliberately inflated in the hope of evading immigration regulations, he believed that victims of persecution were entitled to special dispensation.[100] This assumption, as we shall see, remained with him in later years, although other considerations began to impinge on the scope of the dispensations he was willing to tolerate.

Conditions of immigration into Palestine in general, and immigration from Soviet Russia in particular, began to deteriorate in the second half of 1925, though the deterioration was gradual, unfolding in stages over the next half decade. A number of factors played a role. First, there was a change of personnel in the Palestine Government. In 1925, a new High Commissioner was named to succeed Herbert Samuel. Lord Plumer and his Chief Secretary, G.S. Symes, were not beholden to the Zionist vision of the Balfour Declaration, instead inclined to view their responsibilities in Palestine from a narrow administrative perspective. Under Lord Plumer, government officials who had objected to the Zionist venture from the outset were more free to voice their views, and pro-Zionist officials like Bentwich and Nurock, while still exercising considerable influence over policies that came under their purview, had to be more circumspect.

With regard to immigration, the situation changed in two ways. Legally, the beginning of Plumer's term coincided with the enactment of a new Immigration Ordinance, which had been under discussion for several years. The final version of the ordinance gave the government increased authority to reject or deport 'undesirable' immigrants and to control the flow of certain categories of immigrants.[101] It is doubtful that this change alone would have had a negative impact on immigration, but it was followed in short course by an economic downturn and massive unemployment. The difficulties began in late autumn 1925 and grew steadily worse during the winter and spring, leading the PZE in August 1926 to suspend over two thousand certificates already allocated to individual immigrants and take other measures to limit the number of immigrants.[102] The economic crisis continued unabated in 1927, leaving thousands of immigrants unemployed and causing a negative population flow.

Although the troubles in Palestine caused much discouragement and concern to the Zionist movements in Soviet Russia, the effect on British policies towards immigration from Soviet Russia was still minimal in 1926 and the first half of 1927.[103] The generally sympathetic attitude towards 'political refugees' fleeing the Soviet regime is apparent from letters sent by the government's Immigration Department to the Chargé d'Affaires in Moscow (March 1926) and the PZE (June 1927).[104] By now, such immigration was a matter of routine. The selection of those to be admitted under the Labour Schedule was left entirely to the PZE and, in the case of imprisoned Zionists, to the Political Red Cross (Pompolit) in Moscow. The only restriction concerned the age of Category E immigrants coming under the Labour Schedule (men and women between the ages of 18 and 36). Hyamson urged strict adherence to the rules in this matter as well as the length of validity of the certificates, but he also showed flexibility in adjusting the rules to the special difficulties in communications with Moscow. Until July 1927, Pompolit was supplied every six months with enough blank certificates to meet the needs of all those obtaining permission to substitute their exile for 'deportation'.

Hyamson's role deserves closer attention. He was, after all, the official responsible for the Palestine Government's policy on immigration. In Zionist documents, he is often described as a staunch enemy of Jewish immigration. Moreover, in his post-Palestine years, he was sharply critical of the Zionists, accused them of reckless conduct and argued that, whereas the government took seriously its obligation to regulate immigration in accordance with economic realities, the

PZE allowed itself to be guided by pressure from its supporters and the needs of propaganda.[105] Hyamson's memoirs also reveal the biases that made him take an increasingly negative view of the Zionist settlement of Palestine. The Zionist Organization, he believed, was taken over by 'a party, mostly Russian, rigid in its determination to abide by its principles, without compromise, without diminution by an iota'.[106] Elsewhere, he lamented the role of East European Jews in settling the land. 'The history of Palestine under the Mandate would have been different and probably far happier if Western Jews – especially those of a German, Dutch and British upbringing – had from the beginning had an adequate share in it'.[107] These feelings, though probably unvoiced at the time, poisoned Hyamson's relations with Kisch and the PZE, and in time increased his rigidity in interpreting the rules. Yet he appears to have retained considerable sympathy for those 'truly' persecuted by the hated Soviet regime.

The crisis of 1927 and the last years of substitution immigration

Arguably the greatest damage to the cause of Soviet Zionist immigrants resulted from the deterioration in British–Soviet relations and the Communist scare that followed. The publication in October 1924 of a forged letter purportedly written by G.E. Zinov'ev, head of the Communist International, and containing instructions for subversive activity in Britain, brought down the Labour government of Ramsay McDonald and chilled Britain's budding diplomatic relations with the Soviet Union. Those at the Zionist Executive in London who were in communication with British officials, worried that these events might lead to increased concern about Communist infiltration in Palestine and undermine immigration from Soviet Russia. Thus, Leonard Stein of the Executive's Political Department wrote to Captain J. Jacobs at the PZE, asking for information about the immigrants arriving from Odessa on the Soviet steamer *Chicherin*. Two English tabloids had reported that the boat carried 150 agents of the Communist International, masked as Jewish colonists, and accused the Zionists of allowing the Bolsheviks to make Jaffa into 'the headquarters of the Communist campaign against British rule in Egypt'.[108] Although this story was rife with exaggeration, British apprehensions about the activities of Soviet and Comintern agents in the colonies were confirmed by other, more weighty events (for example, the Comintern's meddling in the revolutionary events in China in 1926).

In practice, immigration from Soviet Russia was not seriously affected until June 1927, when Britain broke diplomatic relations with Moscow, accusing the Soviet Trade Delegation in London of engaging in subversion. The break up of diplomatic relations marked the start of mounting difficulties and a succession of new regulations. From summer 1927 to early 1930 there were no British envoys in the Soviet Union and no possibility of receiving visas for Palestine within its borders. The granting of Immigration Certificates from the Palestine Government was likewise complicated by the heightened fear of Communist infiltration and the absence of a presumably reliable representative in Russia to check into the applicants' particulars.

The weeks and months following the rupture of British–Soviet relations saw an intensive correspondence over the resulting disruption in immigration from Soviet Russia to Palestine and the early efforts to establish new channels. Letters and memoranda were exchanged between the Norwegian Legation in Moscow, which represented Britain's interests; Ekaterina Peshkova, who made urgent presentations to the Legation, requesting a swift solution to the problem of Zionists who had been granted substitution but could not obtain a visa; the Palestine Government in Jerusalem, concerned now above all to ensure against Communist subversion; and the PZE, torn between its continued sympathy and commitment to the Zionist convicts and its anxiety about being linked with a Soviet organization. Two questions dominated these communications: who would have the authority to nominate immigrants, and how would the visa authorizations be transmitted. The latter question was a practical one. Immigrants from the USSR needed to be able to pick up their Immigration Certificates from a British Consul in a location they could reach under the terms of their release from the Soviet penal system. The British Legation in Warsaw offered its services, but the choice fell on Constantinople, a frequent docking stop on the route from Odessa to Jaffa. The matter turned out to be more complicated than at first imagined, because obtaining a Turkish visa to land in Constantinople was exceedingly difficult. By early 1928 it was agreed that immigrants from Russia could pick up their certificates at the point of arrival in Jaffa.

It was, however, the first question that had the more serious consequences for substitution immigration, for it involved a political judgement and was subject to political considerations. Furthermore, it was seriously affected by the fear of Communist infiltration, which ran high in 1927 and mounted higher yet after 1929, when Palestine was swept by a new wave of violence.[109]

At first, attention was focused on the role of Pompolit. Naturally, it no longer seemed advisable to allow an organization based in Moscow to use blank certificates for unnamed immigrants of its own selection. The Zionist Organization and the PZE were as suspicious as the British. The London Executive had expressed serious concern over being compromised when it realized in spring 1927 that a non-Zionist organization in Moscow was playing a central role in arranging the immigration. Meanwhile, the Palestine Office in Warsaw, unaware of these concerns, responded positively when approached by Peshkova and held two meetings with her in April and June 1927 (on her way to Italy and back), after which it turned to the PZE and asked for financial aid for the imprisoned Zionists whose plight Peshkova had so compellingly described. The Executive's response was unusually cautious – it refused to provide the funds and suggested, quite implausibly, that the money could come from 'well-known people' inside Russia. Its reaction to Peshkova's direct contact with the Norwegian Ambassador in Moscow was likewise negative. Anxious to secure its own right to approve all immigrants, the Executive argued that it was imperative to ensure the good standing of every immigrant.[110] Rather than sending money through Peshkova, the PZE discussed the possibility of using the Moscow office of the JDC (which was reported in autumn 1927 to be holding £3,000 to £4,000 for aid to political refugees).

In addition, it sought the help of the Moscow office of the JCA in ensuring the physical suitability of prospective immigrants.[111]

More serious damage to substitution immigration was caused by heightened British suspicions of the Soviet regime and the Comintern. In a letter from 9 August 1927, Hyamson exploited the imprecise terminology used by Peshkova ('collective visas') to deny that the practice of sending her blank certificates had ever existed.[112] Moreover, he now demanded that the names of all individuals seeking Immigration Certificates be submitted to his Section for approval, though he conceded that, even so, 'undesirable elements' could slip into Palestine, a risk that was 'inevitable if Palestine is to be open to Russian political prisoners'.[113] Indeed, the danger of Soviet agents and Communists became a paramount preoccupation of the British for the remaining years of substitution immigration and often outweighed the desire to help critics of the Soviet regime. Thus, in a letter to the Secretary for the Colonies of 5 October 1927, Plumer suggested that to prevent the Soviet government from slipping agents into Palestine under the cover of 'political refugees', all visa applications from such refugees be sent to London for approval – a procedure that had previously applied only to those who were politically suspect. From there, the matter would be returned to the Palestine Government, where the Controller of Immigration would continue to exercise the final authority to grant certificates.[114] In force for the next four years, this procedure greatly slowed down the issuing of Immigration Certificates for Zionist convicts, a situation that was ameliorated only minimally by allowing these immigrants to arrive in Palestine without physical proof of a visa and simply collect their certificates, with photographs attached, at the port of disembarkation.

And yet, the stricter regulations of October 1927 did not signal a decision to stop the immigration of Soviet Zionist convicts. In fact, 'political refugees from the territories of the Union of Soviet Republics' were among a handful of categories exempted from the sweeping restrictions imposed in the second half of 1927 on all immigration to Palestine, and the freeze on the previously approved Labour Schedule, due to massive unemployment.[115] For the next few years, there continued a constant tug between the wish to help refugees from Soviet persecution on the one hand, and a combination of economic considerations and fear of Soviet infiltration on the other. Regulations were tightened, but their implementation remained relatively lax for a while, especially whenever the PZE applied pressure.

Another round of restrictions on immigration into Palestine came in the aftermath of the Arab attacks of 1929, which prompted the British to reconsider the full gamut of their policies in Palestine. With regard to immigration from Soviet Russia, certain concerns were intensified by the increase in Communist activity during the disturbances. Already in the spring of that year, immigrants from Soviet Russia needed a security clearance not only from London, but from the Palestine Police as well. Moreover, guarantees were now required that applicants were 'not connected with the Communist, Left Wing of Po'alei Tziyon or kindred movements', and that their deportation, should they be found to be 'undesirables', would not have to be paid by the Palestine Government.[116] Against all

Zionist expectations, British worries about Communist infiltration only increased when diplomatic relations between Britain and the USSR were renewed in December 1929, since the posting of British consuls to Moscow and Leningrad had made it significantly easier to obtain a visa for Palestine.[117] Hyamson summarized the issues in a memorandum marked, 'Confidential. Immigration from Russia – Political Reliability of Immigrants'. He derided as 'worthless' British reliance on the PZE and on individual residents in Palestine to guarantee the political reliability of candidates. 'In this country... all kinds of certificates are given most light heartedly and merely to do a friend or acquaintance a favour'. Neither did he trust the police and Scotland Yard to identify agents who assumed the identity of particular Zionist prisoners.[118]

For the moment, Hyamson's concerns ran in tandem with the goals of the PZE. The security regulations, so sharply criticized by Hyamson, were seen by the PZE as responsible for unnecessary delays in approving substitution immigrants. Points prepared for Kisch's meeting with Hyamson on 5 January 1930 referred to two lists of 158 political exiles held up by the government for three and four months with no response – in spite of an agreement Kisch had obtained from the Chief Secretariat in early 1928 to speed up the process.[119] On 15 May 1930, Kisch wrote to Hyamson to press for a change in procedure, 'now that a British embassy... has been re-established'. He reminded him that it was 'His Excellency's desire' to expedite the procedure in regard to immigrants from Russia and suggested moving the investigation of prospective immigrants from London to Moscow. Kisch also suggested a way to minimize the number of certificates the PZE had to allocate to the substitution immigration: instead of keeping a certificate for every approved immigrant in Jaffa or Constantinople, a quantity of blank certificates should be sent to Moscow to be filled out by the British Consul, equalling about one-third of the total number of approved immigrants (since many of the applicants were expected to face one or another obstacle to their exit from the Soviet Union).[120]

Beyond the particulars of their parallel attacks on the immigration regulations of October 1927, Kisch shared Hyamson's goal of preventing radical elements from entering Palestine. To be sure, the PZE rejected any British interference in the politics of Zionism. In spring 1929 it had balked at the broader impediments imposed by the Commandant of Police, requiring the Jewish Agency to vouch for the immigrants' political reliability.[121] But it was determined to dispel British fears, and for that reason Kisch agreed after further discussions with the Chief Secretary that the PZE would make every effort to ascertain the party affiliation of those nominated for immigration from Soviet Russia, as long as it did not have to take full responsibility for their political reliability.[122] On this and other occasions, the PZE and the London Executive showed keen sensitivity to British suspicions.[123]

Possible Communist infiltration through the immigration of Zionist convicts was discussed beyond the Palestine Government and the PZE. *Davar*, for example, published an article by a GPU defector, which had appeared in the anti-Bolshevik Russian newspapers based in Paris, *Poslednie novosti* (1931). Claiming to have been the resident GPU officer in Constantinople, he reported intensive work by the GPU and the Comintern in Palestine during and following the violence of 1929. Their agents, he added, entered Palestine 'undercover, as members of Jewish parties opposed to the Soviet regime, who are being deported administratively from the USSR'.[124] In this atmosphere, the PZE implemented its own controls against persons whose political orientation was suspected of excessive radicalism. A review panel was assembled by Kisch and asked to meet regularly to examine the lists of prospective immigrants submitted by the Histadrut's Immigration Centre. Kisch personally promised the panel – composed of the Secretary of the PZE's Immigration Department and representatives of Magen and Illegal Hechalutz – that the objection of any one of them would be sufficient to reject a case, in effect giving the representatives of the non-socialist wing of Soviet Zionism the power to veto any immigrant.[125] Responding to this move, the Immigration Centre rejected any hint of Soviet infiltration of the movements and charged that the new procedure delayed the transmission of lists of prospective immigrants to the British authorities by an additional four to five weeks.[126] Yet it, too, accepted with hardly a protest the government's decision not to admit into Palestine members of the left wing of the socialist party Po'alei Tziyon.[127]

It is against this background of shared fears, that one must understand the surprising cooperation between Hyamson and the PZE during the work of the Interdepartmental Committee on Immigration from Russia to Palestine. Appointed in June 1930 by the new High Commissioner, Sir John Chancellor,[128] and consisting of Hyamson, Nurock and the head of the Criminal Investigation Department, the committee submitted its final report on 30 October 1930.[129] The draft report published here is explicit about the influence of the Zionist Executive. Indeed, on most essential points the Committee's recommendations followed the procedures urged by Kisch a few months earlier: to remove the difficult requirement of sending all applications for Immigration Certificates to London, and to open a British Consulate in Odessa.[130] At the same time, the Committee stood by the requirements that the Zionist Executive vouch in writing for the political good standing of every immigrant on its lists and guarantee funding for possible deportation, should it occur within three years of arrival.

For reasons that are not entirely clear, the new procedures were not implemented for another year. In the meantime, files continued to be sent to London for approval, thus delaying the granting of Immigration Certificates and causing several applicants to miss their opportunity for exit and be returned to exile by the Soviet authorities. In reporting to the Chief Secretary about this situation, Kisch complained bitterly about Hyamson's rigid refusal to help in many cases where the certificates' validity had expired, describing such obstruction as contradictory to the 'traditional attitude of Great Britain towards political refugees in general, and the goodwill that has been both expressed and shown in many cases with regard to Russian Zionists who have been subject to cruel persecution in that country'. The Russian Zionists, he reminded, were persecuted for 'aims and ideals which constitute part of the declared and settled policy of His Majesty's Government in regard to Palestine'. His rhetoric was typical of the Zionist argument throughout.[131]

The final chapter of substitution immigration took place in late 1933 and early 1934. At the time, all concerned believed that new opportunities were about to open up. In Moscow, the British Consul General and Peshkova expected more exit permits to be available, because passport fees had become a source of needed revenue. In Washington, two representatives of Russian Zionism met with Supreme Court Justice Louis Brandeis to lobby for the inclusion of emigration of Zionist exiles in the negotiations on diplomatic relations between the United States and the USSR.[132] In Palestine, a rare meeting of representatives of Magen, the Immigration Centre, the Jewish Agency's Immigration Department, and the Russian Zionist movements took place on 14 December 1933 to discuss procedures, funding and the political context.[133] The questions were as thorny as they had been all along: can *chalutzim* in other countries, suffering their own funding problems, be asked to shoulder the burden of helping the Soviet Zionists? Should there be an appeal for help from non-Zionist Jewish organizations such as the JDC, JCA or Alliance? Should Peshkova be trusted to handle the funds? Will public mobilization cause more harm than good? There was apparent desire to help, but also a sense of helplessness. A few days later, the Director of the Jewish Agency's Immigration Department, Yitzchak Gruenbaum, turned to Hyamson with an urgent request to renew the certificates of several Zionists who were about to be sent back into exile.[134] Hyamson's response to the substance of the request is not known, but three weeks later he reminded the Jewish Agency that it must post a deportation bond for three years for each individual it planned to nominate.[135] Gruenbaum's reaction was angry and reflected the frustrations entailed at this point in the substitution immigration. The requirement was illegal as well as unreasonable.

> There are thousands of applications from Russia, but only a small proportion of them receive permits after long delays and even of these only a fraction is fortunate enough to reach Palestine, in view of the obstacles placed in their way by the Soviet authorities...[136]

But he toned down his objection when he approached Hyamson again on 26 February, asking for immediate action on the case of several Zionist exiles who had been allowed to 'substitute' their penalties for departure.[137] His letter is the latest document we found in the archives which refers to the substitution immigration. This channel of rescue for imprisoned Zionists and their immigration to Palestine – valued highly by most of the actors we had surveyed here, though with certain misgivings by some – was finally blocked by the Soviet government. When conditions again appeared more favourable in 1941, the Jewish Agency and the Histadrut attempted to negotiate the release of Zionist prisoners with the Soviet authorities, but to no avail.[138]

Conclusions

All in all, over a thousand 'substitution' immigrants – probably as many as twelve or thirteen hundred – entered Palestine during the decade from 1924 to 1934.

The flow was uneven; there were high and low points. Although complete figures are unavailable, this much can be established: In the first few years, dozens of immigrants arrived every two weeks on the boats from Odessa. This was when 'substitution' was relatively easy and quick in Soviet Russia, the British government responded enthusiastically and the economy in Palestine provided employment opportunities. The numbers began to decline in 1926 because of increasing obstacles in Soviet Russia and the economic trouble in Palestine. They remained low in 1927 – due to the break up of diplomatic relations and the high unemployment in Palestine – but rose somewhat in the next few years. These are the years for which more reliable figures are available. In his diary, Kisch noted the arrival of 225 'political refugees' in 1927–1928. In notes he left for his successor, he reported another 426 arriving from 1929 and through the first half of 1931. The PZE's Immigration Department reported 456 immigrants from Russia between September 1929 and July 1930, 'most of them political exiles, arrested for the sin of Zionist activity, who had received permission to leave Russia after years of imprisonment and difficult exile'.[139] Finally, in June 1931, the Executive Committee of Magen put the number of substitution immigrants who had arrived by then at more than one thousand.[140]

Another set of figures, likewise incomplete, allows a rough estimate of the percentage of Zionist prisoners freed by substitution. According to one report, there were in June 1931 nearly one thousand Zionists in Soviet prisons and exile.[141] Since the larger share of active Zionists was arrested in the years before 1931, and only a few received substitution after the middle of that year, it may be assumed that the number of Zionist prisoners who for one reason or another remained in prison and exile, unable to benefit from substitution, stood at one thousand plus a couple of hundred. On the basis of these extrapolated figures we can assume that roughly one half of the Zionists sentenced to prison or exile in Soviet Russia left for Palestine under the substitution arrangement.[142]

Those granted substitution owed their exit as much to the work of Pompolit and the organizations and individuals examined in this chapter as to the Soviet authorities. The motives of those working outside the Soviet Union often differed: whereas the British saw the Zionist prisoners first and foremost as proven political opponents of the Soviet regime, the Zionist Organization was motivated by its belief in the convicts' commitment to practical Zionist action. As for the leaders of the Histadrut – especially Ben-Gurion – they found in the socialist wing of Soviet Zionism a perfect fit for their own 'constructivist' mix of socialism and Zionism. Whatever the reason, the Zionist convicts were generally seen as worthy of special assistance. There were, however, various motives and considerations that impinged on the support for substitution immigration in general, and for certain groups and individuals in particular. The 'delegations abroad' and their movements vied for political influence and for the immigration of their members, with the Histadrut and its Immigration Centre favouring the socialist wing over the toilers. The Palestine Executive and the Palestine Government acted out of their fear of Communist and other radical infiltration. And the Zionist Organization failed in raising the money so desperately requested by Peshkova, the exiles and their advocates. At times, these exigencies and concerns delayed the issuance of Immigration Certificates. At others, they prevented convicted Zionists who were already granted substitution from leaving Soviet Russia, causing some of them to be returned to exile.

Notwithstanding the many failures and personal tragedies that were part of the story of the substitution immigration, its very existence enhances our appreciation for the complexity and the opportunities present in the two political realms bridged by this immigration. Soviet Russia of the 1920s, until Stalin was able to gather decisive power during the later part of the decade, was far from the uniform, perfectly hierarchical political system that we sometimes imagine, allowing for considerable personal and organizational initiative at the highest reaches of the party, soviet and governmental structures. Likewise, in British Palestine, Zionist Executive leaders often worked in close cooperation with their presumed opponents in the Palestine government and, at the same time, shared many of the goals of the labour movements organized in the Histadrut. Indeed, while the substitution immigration is a unique and surprising chapter in the histories of both Soviet Russia and of the Jewish settlement in Palestine – albeit a short and limited one – the relations underlying it were not exceptional but, rather, indicative.

Notes

1 Lavon Archive, IV-211-1, folder (f.) 35.
2 See, for example, Ben-Gurion's letter to Yisrael Idelson (Bar Yehuda), in which the head of the Histadrut was called to explain to the leader of TzS why he had been praising the socialist wing of Ha-Shomer ha-Tza'ir, leaving the youth movement of TzS largely unmentioned. Ben-Gurion explained that he had been highly impressed with the former during his visit to Moscow, while he never met the people of TzS Yugend. *Igrot David Ben-Gurion* (Tel Aviv: Am Oved and Tel Aviv University, 1973), Vol. 2, p. 253.
3 In the early twentieth century, the Social Democrats, Socialist Revolutionaries (SRs), and even the Constitutional Democrats established 'delegations abroad' (in Russian *zagranichnye delegatsii*). The Mensheviks and SRs revived their delegations when legal opposition to Bolshevik rule became impossible in Russia.
4 See, for example, the report sent from Moscow on 21 November 1924, listing recent sentences passed against members of TzS: Yad Tabenkin, section 15-Marom, series 2, box 4, f. 1.
5 Circular to all municipal Workers' Councils and agricultural settlements, 27 February 1925. Lavon Archive, IV-211-1, f. 34.
6 Letter of L. Dobkin to the Immigration Department, PZE, 28 April 1925, Lavon Archive, IV-211-3, f. 17A. For Tze'irei Tziyon, the address given was PolitKrest, for TzS – Vilensky at Gazetnyi 3/21.
7 Letter from Chaim Halperin to Avraham Arest, 27 January 1928, Lavon Archive, IV-211-1, f. 17, seeking information from former members of Zionist movements for the Zionist Executive in London.
8 About the ambivalence of the labour leaders in Palestine in relation to Soviet Russia, see Anita Shapira, *Ha-Halikha al Kav ha-Ofek* (Tel Aviv: Am Oved, 1989), pp. 118–56 and 258–92. See also the correspondence between Ben-Gurion and Israel Idelson (Bar Yehuda): *Igrot Ben-Gurion*, Vol. 2, p. 285.
9 *Igrot Ben-Gurion*, Vol. 2, 164–5, 170–2.

10 Immigrants from three non-socialist youth movements (Hechaver, Kadima and Histadrut ha-No'ar ha-Tziyoni) met as early as October 1924 and called on the organized workers in Palestine to support their comrades in Soviet Russia. *Kuntres*, Vol. 9, No. 192, 7 November 1924, p. 28. A few weeks later, TzS issued its call for help 'To the Jewish Workers All Over the World'. *Pinkas*, Vol. 3, Nos 4–5, 8 January 1925, pp. 137–8.
11 *Kuntres*, Vol. 11, No. 214, 1 May 1925, p. 40.
12 *Kuntres*, Vol. 11, No. 215, 8 May 1925; Vol. 11, No. 216, 15 May 1925, p. 40.
13 One example is the meeting in January 1928, coordinated by the Executive Committee of the Histadrut and attended by activists of the socialist movements, among them A. Tzizling and El. Iskoz (Galili). Yad Tabenkin, section 15-Galili, box 45, f. 2.
14 For example, the protest resolution adopted by the Third Conference of the Histadrut in July 1927, which occasioned a walkout by the party of Left Po'alei Tziyon and the Left Bloc. *Davar*, 11 July 1927.
15 *Kuntres*, Vol. 17, No. 334, 20 April 1928, pp. 28–9; *Davar*, 22, 23, 25, 28 March, 18 and 23 April and 6 May 1928. Two years earlier, *Davar* had dedicated an editorial to the first attack on Legal Hechalutz, when Pines and others were arrested in March 1926, *Davar* 29 March 1926.
16 For example, reports on a meeting between *Chicherin* and a delegation of German Jews (*Davar*, 20 October 1925) and a Evsektsiia memorandum regarding Zionism (*Davar*, 28 October 1925); and discussion of interviews with Lunacharskii (*Davar*, 22 July, 9 and 20 September and 9 October 1928).
17 *Davar*, 2 August and 27 December 1925 (Ha-Shomer ha-Tza'ir); 28 March and 21 June 1926 (Hitachdut); 26 March and 6 May 1926 (Legal Hechalutz); 22 October 1928 (Illegal Hechalutz).
18 V. Kostrinskii, 'On the Condition of Exiled Zionists in Russia', *Davar*, 19 August 1926; 'From the Prison in USSR', 16 February 1927 (based on information from the 'Delegations Abroad' of TzS); 'From the Tortures of our Comrades in Russian Exile', 18 April 1928; 'A New *Gzeira* on the Prisoners of Zion in Russia', 23 November 1927.
19 *Davar*, 6 August 1925 (joint protest meeting in Tel Aviv); 13 August 1925 (Appeal of the moderate labour party Ha-Po'el ha-Tza'ir); 8 May 1927 (announcement of the Organization of High School Students and Youth Organizations to Help the Prisoners of Zion).
20 *Davar*, 22 May 1927, p. 3. At the same time, Eliyahu Dobkin, representing World Hechalutz, went to London and Paris to organize similar committees. In London, Dobkin found that Hitachdut had collected funds on its own and was reluctant to join the effort, but in the end the London committee included its representatives as well as Hechalutz, TzS, and Ha-Shomer ha-Tza'ir. Dobkin's letter to Tzizling, dated 18 January 1928. Yad Tabenkin, AKM, Chativat Chul, box 2, f. 7, document 25.
21 EVOSM's memorandum to the fourteenth Congress, 'El ha-Tzirim ve-ha-Orchim shel ha-Kongress ha-Tziyoni', Tel Aviv, 1925; the appeal of the United Committee to Help the Prisoners of Zion to the fifteenth Congress, *Davar*, 28 August 1927, p. 3.
22 The meeting took place on 25 May 1929, chaired by the poet Chaim Nachman Bialik. *Magen: Kovetz mukdash le-she'elot ha-Tziyonut ha-nirdefet be-Rusya ha-Sovyetit* (Tel Aviv: Magen, June 1931). pp. 19–21; *Davar*, 26–7 May 1929; interview with Binyamin Vest, 1974. Yad Tabenkin, T/25, box 10, f. 3, p. 18.
23 Goldberg had worked in the area of Hebrew culture in Moscow. He was arrested three times before being allowed to leave on the *Novorossiisk* in July 1924.
24 Letter dated 5 October 1924. Lavon Archive, IV-104-53, f. 35.
25 Israel Rosoff to Colonel Kisch, Chair of PZE. Tel Aviv, 11 March 1927. Central Zionist Archives (CZA) Jerusalem, S25/10259. See also letter from the United Committee to Help Prisoners of Zion in Soviet Russia to Weizmann, Jerusalem, 12 October 1927. CZA S25/2426.

26 CZA S25/657.
27 Riga, 8 March 1927, marked 'Secret'. CZA Lg/285.
28 These wings coalesced around the parties of the TzS and STP.
29 Riga, 23 November 1924. CZA Z4/1736. For Zinov'ev, see p. 59 in this volume.
30 Main Bureau, World League of TzS, Danzig, to Zionist Executive Committee, London, 16 November 1924. CZA Z4/1736 and Lavon Archive, IV-211-3, f. 2. The same accusations are voiced in the unpublished letter of Yisrael Idelson to Yisrael Mereminsky, Danzig, 28 August 1924, Yad Tabenkin, section 15-Marom, series 2, box 4, f. 1. See also World Hechalutz to the Immigration Centre, 11 November 1924. Lavon Archive, IV-211-3, f. 2.
31 Zionist Executive, London to F.H. Kisch, 2 December 1924; and protocol of PZE meeting on 14 December 1924 (the speakers for this position were Shkolnik and Kaplansky). CZA S25/650.
32 Menachem Rivlin, 'Be-tzel ha-Ye'ush' (In the Shadow of Despair). Reprinted in *Menachem Rivlin, Bein Sifrut le-Sapanut* (Haifa: Zim Shipping Company, 1986), pp. 53–6.
33 M. Zeltzer, 'About A Forgotten Duty', *Davar*, 18 January 1927. See also M. Zilist, 'Immediate Help Is Needed!' *Davar*, 29 December 1926; Alef, 'Hurry With Help', *Davar*, 25 December 1927.
34 A history of the Immigration Centre and its impact on immigration to Palestine has yet to be written. The discussion here is based on documentation preserved in the Centre's archive at Lavon Archive and the memoirs of Tzvi Livne (Liberman), *Pirkei ha-Aliya ha-Shlishit mi-Knisat ha-Britim la-Arets ad Ve'idat ha-Histadrut* (Tel Aviv: Sifrei Gadish, 1958). Livne headed a precursor of the Centre out of the offices of his party, Ha-Po'el ha-Tza'ir.
35 In its first year of work, the Immigration Centre established and ran a dormitory, dining hall, laundry and postal centre for the penniless immigrants. Most of these services were taken over in 1920 by the PZE's newly created Immigration Department under the leadership of Yehoshu'a Gordon. Livne, *Pirkei ha-Aliya ha-Shlishit*, pp. 75–80.
36 Immigration Centre to the World Centre of Hechalutz, 29 October 1924, Lavon Archive, IV-211-2. For a discussion of the political issues surrounding immigration to Palestine in those years, and the role of Hechalutz, see Moshe Lisak, '*Aliya, Klita u-Binyan Chevra Yehudit be-Eretz-Yisrael bi-Shnot ha-Esrim* (1918–1930)', in Moshe Lisak, Anita Shapira and Gavriel Cohen (eds), *Toldot ha-Yishuv ha-Yehudi be-Eretz-Yisrael me'az ha-Aliya ha-Rishona*, Part Two, *Tkufat ha-Mandat ha-Briti* (Jerusalem: Mosad Bialik, 1995).
37 A letter from Mereminsky to the PZE from autumn 1924 explains that information about six members of Ha-Shomer ha-Tza'ir sentenced to exile and in need of visas was sent by 'the *responsible* person who had been appointed by us to handle these matters for Ha-Shomer ha-Tza'ir'. Lavon Archive, IV-211, f. 17A.
38 From the protocol of a meeting of the Immigration Centre on 27 February 1930. Lavon Archive, IV-211-1, f. 37.
39 Yisrael Mereminsky was among the founders of the Tze'irei Tziyon party in 1917 and later its representative to the Jewish Assembly in Ukraine. He left for Warsaw, where he was active in TzS-Tze'irei Tziyon until his immigration in 1923. Levi Dobkin was an activist of Hechaver and later a member of the Central Committee of Hechalutz. He was arrested in Moscow in March 1924, exiled to the Chuvash region, but allowed to exchange his term there for departure to Palestine in September 1924. (See letter from Hechalutz Centre to R.P. Katanian, General Procurator, dated 10 June 1924, GARF, f. 7747, op. 1, d. 16, l. 109.) Chaim Halperin, who had joined Hechalutz in its first year, was a member of the Central Committee of TzS, and remained with the 'legal' wing after the split.

40 Lavon Yisrael Archive, section IV-211-3, f. 2, pp. 65–7. The meeting took place earlier that day. Yisrael Mereminsky initialled the letter.
41 The letter is dated 11 November 1924. Lavon Archive, IV-211-3. f. 2, pp. 48–9.
42 The letter to the Immigration and the Political Departments, dated 28 April 1925, signed by L. Dobkin. Lavon Archive. IV-211-3, f. 17A.
43 See Document 5 in this volume: Y. Mereminsky, Immigration Centre to Central Office of World Hechalutz (Berlin).
44 Lavon Archive, section IV-211-3, f. 2, pp. 4–6.
45 Lavon Archive, IV-104-53, f. 46; Ben-Gurion, *Igrot Ben-Gurion*, Vol. 2, p. 282. Vest even declined to corroborate lists of imprisoned members of Illegal Hechalutz which the Immigration Centre submitted to the Immigration Department. Letter to the PZE. 8 March 1925, signed by L[evi] D[obkin] for Ben Tzvi. Lavon Archive, IV-211-1. f. 17A.
46 I. Kostrinski, 'Bias', *Davar*, 13 September 1926; 'A Meeting of Members of Illegal Hechalutz', *Davar*, 19 June 1928.
47 Ben-Gurion to Eliyahu Golomb, 17 November 1924, *Igrot Ben-Gurion*, Vol. 2, pp. 255–8; 'The Question of "Hechalutz" at the Danzig Conference', *Davar*, 11 May 1926.
48 Lavon Archive, IV-211-3, f. 2, pp. 4–6.
49 The exchange between Nachum Verlinsky and Yakov Rabinovich about Jewish settlement in Crimea, carried on the pages of *Ha-Po'el ha-Tza'ir* and *Davar*, 13 September 1925, is one example of the continuing division between the two wings of Soviet Zionism.
50 Letter to PZE, 25 November 1926, Lavon Archive, IV-211-3. f. 18B.
51 His letter to L. Motzkin, Paris, 8 May 1928. CZA Lg/285.
52 The Immigration Centre to PZE, 15, 16 July, 24 November 1925, 13 January, 31 August, 25 November 1926, 9 November 1927; the Immigration Centre to Dan Pines, Hechalutz in Moscow, autumn 1925 through 1926, Lavon Archive. IV-211-3. ff. 17B, 18 and IV-211-3, ff. 2, 2A; N. Verlinsky for the Central Committee of Hitachdut to the Zionist Executive, London, 8 March, 1927, CZA Lg/285.
53 Yakov Kostrinskii, 'To Ransom the Prisoners', *Davar*, 9 March 1927.
54 El. Galili (for 'Kibbutz Ha-Shomer ha-Tza'ir from the USSR'), 'About One Saddening Matter', *Davar*, 24 March 1927; B. Vest, 'New Restrictions', *Davar*. 20 November 1927; Berl Reptor, 'Hechalutz and Immigration', *Kuntres*, Vol. 16, No. 330, 29 February 1928, pp. 11–14; Ch. Halperin, 'Upon the Renewal of Immigration', *Kuntres*, Vol. 17, No. 338, 1 June 1928.
55 For discussion of the concerns of the Zionist Organization during these years. see Yigal Elam, *Ha-Sokhnut ha-Yehudit. Shanim rishonot. 1919–1931* (Jerusalem: Ha-Sifriya ha-Tziyonit, 1990).
56 Meetings of 2 February, 29 May, 4, 13 and 20 June 1924. CZA S25/656. On 2 February, the Executive formed a two-member sub-committee to handle the issue of Soviet Zionists, comprising of Drs Halpern and Jacobson.
57 CZA S25/657.
58 For example, the demands presented in 1926 by Vilenchuk and Verlinsky, or those presented by David Ben-Gurion to Weizmann for visas and £5,000 for the immigration of 500 *chalutzim* reportedly approved by the Soviet authorities. Ben-Gurion's meeting with Weizmann took place on 27 July 1924. David Ben-Gurion, *Zikhronot* (Tel Aviv: Am Oved, 1971), p. 285.
59 CZA S25/657.
60 See Document 2 in this volume: Memorandum of A. Merezhin, Central Bureau of the Evsektsiia to the Politburo, and note 7 there.
61 The Jewish Emigration Society was established in 1909 to regulate the massive emigration of Russian Jews to the United States and direct able bodied emigrants to areas where work was deemed more readily available. Its operations were greatly

reduced with the start of WWI, but it maintained a presence in Soviet Russia into the 1930s.
62. Halperin to L. Motzkin, Paris, 8 May 1928. CZA Lg/285.
63. See Document 18 in this volume: Central Office of the Zionist Organization (London) to Members of the Executive Committee and to the Zionist Federations and Fractions.
64. Aviva Chalamish, 'Mediniut ha-Aliya ve-ha-Klita shel ha-Histadrut ha-Tziyonit, 1931–1937'. Doctoral thesis, Tel Aviv University, 1995, p. 260. At precisely the same time, large funds were being gathered to ease the absorption of German Jews in Palestine, with each German immigrant receiving, on average, 100 Palestinian pounds – equal to the cost of a Soviet passport. Ibid., p. 259.
65. Eilam, *Ha-Sokhnut ha-Yehudit*, pp. 80–1.
66. See his letters to Frederick H. Kisch (11 October 1925) and to the American Jewish leader Louis Marshall (12 October 1925 and 13 January 1927), *Letters and Papers of Chaim Weizmann*, Vol. XII-Series A, Joshua Freundlich (ed.) (London: Oxford University Press, 1977), pp. 415–17, 423–5; Vol. XIII-Series, Pinhas Ofer (ed.) (London: Oxford University Press, 1977), pp. 188–90.
67. New York, 21 November 1928. *Letters and Papers of Chaim Weizmann*, Vol. XIII-Series A, p. 516.
68. London, 14 May 1930. *Letters and Papers of Chaim Weizman*, Vol. XIV-Series A, Camillo Dresner (ed.) (London: Oxford University Press, 1978), p. 285. Warburg who had been a founding chair of the JDC, served at the time as chairman of the Administrative Committee of the expanded Jewish Agency.
69. Virulent attacks on the Zionists in Soviet Russia were aired at that time by the writer Chaim Breinin, a veteran Zionist, who visited Russia in 1926 and 1930 and came back denying Soviet persecution of Zionism.
70. CZA S25/2426.
71. *Letters and Papers of Chaim Weizman*, Vol. XIV-Series A, pp. 137, 174–5. Reporting on these contacts to the American Zionist leader Stephen Wise, he added that 'in the rough and tumble of a terrible battle in which we are engaged now we haven't forgotten this particular aspect of our movement'. Ibid., pp. 180–1.
72. See Document 5 in this volume: Y. Mereminsky, Immigration Centre to Central Office of World Hechalutz (Berlin), and note 42 there.
73. On 15 August 1924, Kisch wrote to Stein complaining about an article in a British paper that created the impression of Legal Hechalutz as a 'Bolshevik' organization. 'As you know, I have for months been engaged in persuading the Government here that the Hechalutz exists with difficulty in Russia, that it has no track with the Soviet Government, that its members are the best possible immigrants for Palestine'. CZA S25/650.
74. The division of responsibilities is easily apparent in the PZE's documents. Chalamish discusses it in the context of the 1930s.
75. CZA Z4/1246 II, 1246 III; S6/5295; S6/5350; S25/605.
76. In addition, the meeting called on all the Palestine Offices in Europe to give priority to prospective immigrants trained by Hechalutz 'and other elements educated in the spirit of productive labour'. The meeting's results were reported in a letter from Chaim Pick to Dr Halpern at the London Executive, 3 April 1925. CZA S6/5350.
77. CZA S6/5350.
78. Chaim Pick to Yosef Shprintzak, 30 March 1927. CZA S25/609/9. Figures cited by Aviva Chalamish show that the PZE continued to give priority to Soviet Zionists during 1931, when the number of certificates remained very low: 75 out of the total of 500 in the Schedule for the period April through September 1931; and an additional 75 (for 'exiles of Zion' already approved for immigration) out of 350 certificates in the Schedule for the next six months. Ironically, just when the number of certificates began to increase dramatically (to 5,500 every six months in 1933), the number of

requests from Russia dropped to two or three dozens. The data for 1934 and beyond does not list any allocations for immigrants from Soviet Russia. Chalamish, 'Mediniut ha-Aliya ve-ha-Klita shel ha-Histadrut ha-Tziyonit', pp. 404–16.
79 For a discussion of the division of authority within the PZE in all matters pertaining to immigration, see Chalamish, 'Mediniut ha-Aliya ve-ha-Klita shel ha-Histadrut ha-Tziyonit', pp. 52–4.
80 Lt Colonel F.H. Kisch, *Palestine Diary* (London: Victor Golancz, 1938), pp. 127, 134.
81 F.H. K[isch] to Chief Secretary, Jerusalem, 2 March 1924. CZA Z4/1246/II.
82 Kisch, *Palestine Diary*, pp. 138–9. Kisch mistakenly names him Jesaiah.
83 Jerusalem, 9 July 1924. CZA S25/746.
84 CZA S25/622. See also Kisch, *Palestine Diary*, p. 134, for his 26 June conversation with Ben-Gurion on this issue.
85 Jerusalem, 6 May 1925. CZA S25/746. The request was made in the context of a report on his conversation with Shkolnik and Remez who had just returned from a visit to Moscow.
86 Internal PZE memorandum, Ariav to Bekhar, 5 January 1930. CZA S25/2425.
87 His usual contact there was Assistant Secretary Shuckburgh. Kisch to Naamani, London, 15 October 1927. CZA S25/2426.
88 Notes on immigration from Soviet Russia, left by Kisch for his successor (1931). CZA S25/2426.
89 Dan Pines writes about negotiations in this matter, held by Ben-Gurion with the Foreign and Colonial Offices. He claims to have found out about these negotiations when the NKVD, making reference to a query received from the Commissariat of Foreign Affairs, asked him whether Hechalutz would be able to send 500 trained *chalutzim*. Dan Pines, *Hechalutz be-Kur ha-Mahapekha. Korot Histadrut Hechalutz be-Rusya* (Tel Aviv: Davar, 1938), p. 233.
90 Public Record Office (PRO), London, FO 371, f. 10497.
91 Letters of F.H. K[isch] to Chief Secretary, Jerusalem, 2 March 1924. CZA Z4/1246/II. Max Nurock for the Chief Secretary to the Chairman of the Palestine Zionist Executive, 25 June 1924. Israel State Archives (ISA), Jerusalem ISA 11/1174, IMM/7 I.
92 See his letters from 1924: *Igrot Ben-Gurion*, Vol. 2, pp. 199–200, 231–2, 236–7, 240–1, 255, 260. About his hopes for cooperation with Soviet institutions, see his letters from 1923, Ibid., pp. 168–9, 181–2, 214.
93 PRO FO371/11022. Copies of the correspondence leading up to the explanation are in PRO, FO371, f. 10497. Interestingly, the Moscow Mission also reported that the Soviet Trade Delegation in London had indicated to the Zionist Organization and the Federation of Jewish Workers that Ben-Gurion and Shkolnik should use their proposed visit to Moscow to conclude an agreement about the immigration of *chalutzim*: PRO, F371, f. 11022, pp. 226–7. See also letter sent from the Colonial Office to the Zionist Organization, 2 April 1925, pp. 228–31. As we know from the testimony of David Shor, at this very time, in spring 1925, he had received assurances from Kamenev and Smidovich of their support for large-scale immigration to Palestine.
94 Kisch to Hyamson, 11 July 1924, a note accompanying a report in the *Manchester Guardian* on the departure of forty-six Zionists for Palestine on the *Novorossiisk*. Kisch thanked Hyamson for 'the promptness with which you assisted us in this matter'. ISA 11/1174, IMM/7 I.
95 ISA 11/1177, IMM/7 I. That a consul was never posted had more to do with the state of Soviet–British relations than the immigration of Soviet Jews.
96 M. Mossek, *Palestine Immigration Policy under Sir Herbert Samuel: British, Zionist, and Arab* Attitudes (London: Frank Cass, 1978), pp. 38–9.
97 The Agricultural Centre of the Histadrut obtained a consignment from the government to handle all tobacco harvesting and packing as well as permission to bring as many as 500 immigrants for this purpose, a situation that occasioned unusual criticism from

the Immigration Centre in Tel Aviv against Legal Hechalutz in Moscow for failing to send information about its members who could be recruited for the work. The Histadrut, it was said, could face a crisis and be undermined 'by the cheap labour of workers who have no class consciousness and no cultural needs'. The writer stressed that 'had there been greater immigration of *chalutzim* from Russia, we would not have faced this whole problem'. Signed by Mereminsky, 9 November 1924. Lavon Archive, IV-211-3, f. 2, pp. 4–6.

98 Bernard Wasserstein, *The British in Palestine: the Mandatory Government and the Arab-Jewish Conflict 1917–1929* (London: Basil Blackwell, 1991), pp. 91, 208–19; Mossek, *Palestine Immigration Policy under Sir Herbert Samuel*, pp. 7–8. Other prominent Anglo-Jews included Colonel Harold Solomon (Controller of Stores), Dennis Cohen and N.I. Mindel (officials of the Immigration Department) and Edwin Samuel, the son of Herbert Samuel who held several positions throughout the Mandatory years. See also Tom Segev, *One Palestine, Complete: Jews and Arabs under the British Mandate* (London: Abacus, 2001).

99 Kisch, *Palestine Diary*, pp. 127, 142.

100 Dated 16 September 1925. ISA 11/1147, CONS/E/37/2.

101 The latter was achieved by breaking down the five main categories into many sub-categories. Mossek, *Palestine Immigration Policy under Sir Herbert Samuel*, pp. 95–114.

102 Ibid., pp. 130–3.

103 Kisch reported to the London Executive about his meeting on 14 September 1925 with Plumer, Hyamson and Bentwich, to discuss the new Immigration Ordinance. Most of his objections were rejected outright, but Plumer seemed earnestly interested in Kisch's request to post a British Immigration Officer to Odessa, in order to ease Jewish migration from Soviet Russia to Palestine. CZA S25/651.

104 Letters dated 12 March 1926 and 30 June 1927. ISA 11/1147, CONS/E/37/2 and IMM/7 I.

105 A.M. Hyamson, *Palestine Under the Mandate* (London: Methuen & Co., 1950), p. 56. In a letter to the Chief Secretary, dated 2 September 1927, Hyamson lashes out at the PZE for treating every immigrant who arrives without a visa as an emergency case – 'The PZE will refrain from guaranteeing no one' – thus encouraging unauthorized immigration: ISA IMM/7 I.

106 Hyamson, *Palestine Under the Mandate*, p. 115.

107 Ibid., p. 59.

108 The PZE responded that most of those arriving on the *Chicherin* were 'nominees of this office [who] had been sentenced to deportation to Siberia and were released on receipt of permits from the Palestine Government'. Letters dated 9 and 18 February 1925. CZA S25/651.

109 British reports from summer 1930 on Communist activity in Palestine. ISA (microfilms), FO 371/14500 (f. 734).

110 CZA S25/2424, 2425. The PZE also emphasized the obligation it had to relatives and friends already in Palestine.

111 Internal PZE memorandum, Ariav to Kisch, 11 November 1928. CZA S25/2425. Ariav served then as General Secretary of the PZE.

112 A.M. Hyamson to Chief Secretary, Jerusalem, 9 August 1927, marked Secret. ISA 11/1174, IMM/7 I.

113 In a letter dated 2 September 1927, again to the Chief Secretary, Hyamson reported that since two immigrants arriving without visas had been admitted in May of that year (on assurances from Naamani of the PZE, who identified one of the immigrants, Eliahu Seidler, as a Zionist who had escaped from prison), 'every boat from Russia for months has arrived with people without visas for Palestine'. ISA 11/1174, IMM/7 I.

114 See Document 14 in this volume: H.O. Plumer, High Commisnier for Palestine to L. Amery, Secretary of State for the Colonies.

115 The exemption doubtless resulted from strenuous lobbying by the PZE, as is apparent from the Chief Secretary's letter to the PZE, dated 3 October 1927, which reported the quick approval of three lists of prospective immigrants from Soviet Russia (all submitted in the first half of September): 110 of the 121 names submitted under the Labour Schedule, and 35 (out of 37) submitted in the class of 'Jewish political refugees'. ISA 11/1174, IMM/7 I.
116 See Document 19 in this volume: A. Saunders, Department of Police & Prisons, Government of Palestine to Chief Immigration Officer. In summer 1928, tourists were subjected to the stricter regulations applied to immigrants in force beginning in autumn 1927, a measure supported by the argument that 'the geographical position of Palestine makes it an excellent distributing centre for Bolshevist agents and propaganda'. PRO, CO 733/159/16. For Po'alei Tziyon see note 127.
117 The procedure for obtaining Immigration Certificates continued to be very complicated, but once a certificate was authorized, the consuls were notified and could issue a visa directly to the individual. See Document 21 in this volume: Draft Report of Interdepartmental Committee, Government of Palestine on Immigration from Russia.
118 Dated 4 March 1930. ISA 11/1174, IMM/7/8.
119 CZA S25/2425. To be sure, disagreements remained, not least of them entrust Hyamson's proposal to entrust the political screening of prospective immigrants to the Moscow office of the JDC (which had become in 1929 a constitutional part of the Jewish Agency and led its non-Zionist wing).
120 ISA 11/1174, IMM/7 II.
121 Kisch to Hyamson, 7 May 1929. ISA 11/1174, IMM/7 II.
122 Chief Secretary to PZE, 4 June 1929. ISA 11/1174, IMM/7 II; list of prospective immigrants sent by PZE to Chief Secretary, 10 June 1929: CZA S6/5355. See Document 20 in this volume: Palestine Zionist Executive to the Chief Secretary, Government of Palestine.
123 See, for example, report on the meeting of L.J. Stein and Dr Brodetsky with Sir John Shuckburgh, Colonial Office, on 22 July 1929. CZA S25/2426. In a letter dated 19 February 1928, Halperin wrote to Barlas at the PZE's Immigration Department to respond to British suspicion that four recent immigrants were Communists. Lavon Archive, IV-211-1 f. 23, pp. 57–8. In a personal note he added, ('so that there will be no doubt in your heart'), that none of the four immigrants in question had belonged to his own party (that is, Legal Hechalutz): 'they are from the "Right" in the labour movement – EVOSM, Hitachdut, and Illegal Hechalutz'. Lavon Archive, IV-211-1, f. 23, pp. 57–8.
124 Lavon Archive, IV-104-53, f. 164.
125 Kisch's notes to his successor (1931). CZA S25/2426.
126 At the meeting of the Immigration Centre on 27 October 1929, Chaim Halperin explained that Kisch entrusted the review to the General Zionist Israel Rosoff. Lavon Archive, IV-211-1, f. 37. But in March 1930, Barlas reported on a meeting to review Peshkova's lists of prospective immigrants, which was attended by Chaim Halperin, Miriam Shtarkman (Illegal Hechalutz), Shne'ur Aronoff (TzS), and Lang (Magen). CZA S25/2426.
127 Po'alei Tziyon, a Social Democratic Zionist party, was founded by Ber Borochov in Russia in the early 1920s and became a worldwide movement. After 1917 the movement in Russia split between those who renounced Zionism and joined the Communist Party, and a faction that supported both Jewish settlement in Palestine and Communism (called the Evreiskaia Kommunisticheskaia Partiia (EKP), that is, the Jewish Communist Party). The world organization was split in 1920 between a right wing that affiliated with the World Zionist Organization, and Left Po'alei Tziyon, which rejected the WZO and sought admission to the Communist International.

128 Plumer ended his term in mid-1929 but Chancellor arrived only in December 1929. During the interregnum, things were run by Chief Secretary C.H. Look.
129 See the preamble to the Committee's final report. ISA 11/1174/2335, IMM/7/8.
130 The Committee was also supplied with a translation of several bulletins on the situation of Jews and Zionists in Soviet Russia, compiled by the 'delegations abroad' of Hitachdut and TzS. ISA 11/1174, IMM/7/8.
131 F.H. K[isch] to the Chief Secretary, Jerusalem, 14 April 1931. ISA 11/1174, IMM/7/5/1.
132 The meeting with Justice Brandeis was attended by Chaim Grinberg, a well-known leader of the Zionist Organization in Russia, and Elazar (Lasia) Galili, representing Ha-Shomer ha-Tza'ir and other movements with exiled and imprisoned members and took place on 14 November 1934, during Litvinov's visit to Washington. Letter of Lasia (Elazar) Galili to Klara Galili, 13 November 1934, Baltimore; Elazar (Lasia) Galili, *Zikhronotav ve-Kavim li-Dmuto*, Afikim 1997. These two sources are in the author's possession.
133 CZA S25/2427.
134 Dated 22 December 1933. ISA 11/1174, IMM/7/6 II.
135 Dated 15 January 1934. ISA 11/1174, IMM/7/6 II.
136 Dated 7 February 1934. ISA 11/1174, IMM/7/9.
137 See Document 22 in this volume: Y. Gruenbaum, Jewish Agency to Director of Immigration, Government of Palestine, Jerusalem.
138 CZA Z4/14920.
139 A more detailed Immigration Department report, covering the period September 1929 through February 1930, explained that a large share of the certificates as well as a special budget of £1,000 were allocated for victims of persecution in Russia, allowing 250 immigrants to come, of which 124 were political exiles. CZA S6/5454.
140 *Magen: Kovetz mukdash le-she'elot ha-Tziyonut ha-nirdefet be-Rusya ha-Sovietit*, p. 3.
141 Arye Tzentziper, 'A Harmful Weapon in the Hands of Those Who Hate Us (by way of response to Breinin)' in ibid., pp. 14–18.
142 In August 1928, the Palestine-based Committee to Help Persecuted Russian Zionists, reported that 980 Zionists were then serving terms in prison and exile. Forty per cent of them had been given eligibility for substitution. *Davar*, 2 August 1927.

4 Postscript
Two fates

Ziva Galili

In 1990, as the Soviet Union was disintegrating, two elderly sisters travelled from Moscow for a visit in Israel. Raisa (Raia) and Bella (Badia) Borovaia had spent a good part of their lives in exile, finally settling after Second World War in a small village in Kazakhstan, where they worked as factory hands until the middle of the 1980s.[1] In their childhood and youth, they had lived in Moscow with their well-to-do parents and two other sisters – Fira, the eldest, and Fania, who was Raia's twin. In 1919 or 1920, their father enrolled all four girls in the Jewish sports organization Maccabi. From there they went on to become members of the youth movement Ha-Shomer ha-Tza'ir. Fira was arrested in 1925 and allowed to leave immediately for Palestine. Raia and Fania were arrested three years later. Both were granted substitution but decided that only Fania should leave: she went on to live in Kibbutz Kfar Gil'adi in the Upper Galilee. Raia stayed behind, convinced that experienced cadres were needed in Russia to keep the movement going. She escaped from exile in 1931 and worked underground, settling down in Gorkii, where she had a daughter with Genrikh Driker, another Zionist escapee from exile. Both were member of the 'command' of Ha-Shomer ha-Tza'ir until their final arrest in 1935. Long terms in political prison followed.[2] This time, the youngest sister, Badia, was also arrested, although she had only been connected to the movement for a brief time in her childhood.

The arrival of the two long-lost sisters was a remarkable event for their siblings and old comrades in Israel, but it was not an entirely happy celebration. There were the long years Raia and Badia had lost in exile and camp. And there was the painful discovery that life had opened a rift between the members of this once close group. Those whose lives were intertwined with the history of the Jewish settlement in Palestine were now estranged from the sisters whose life experience was defined by Soviet dictatorship. Both sides knew hard work, disease, death and suffering. But the settlers in Palestine understood their hardships as serving the purpose they had all embraced in their youth – the building of a new and socially just Jewish society – whereas their comrades in the USSR perceived their own suffering as a terrible waste. It was, for the most part, the arbitrariness of the 'substitution' system that had sent each to a different fate.

This Postscript will sketch out the life trajectories of the Zionist exiles and of the substitution emigrants. First, however, it must be noted that not all the many thousands of young people, who were active in the Zionist organizations of the 1920s, faced the prospect of either prison or emigration. Many escaped arrest altogether, while others were released almost immediately because of their tender age. Little is known about these people, but it can be assumed that most drifted away from Zionist activity and, like many other Soviet Jews, joined the great social mobilization that began in the late 1920s. Many probably entered educational institutions, mostly technical, and worked in some part of the Soviet industrial system or the bureaucracy. The men amongst them surely fought in the Second World War, and many perished, as did many of their families in the western regions of the Soviet Union.

The Zionists who did spend time in prison and exile, and who could not obtain substitution, attempted to build a new life once their terms had ended, though they were restricted to reside in the 'minus' towns, far from the big cities and concentrations of Jewish population. Many were rearrested during the Great Terror of 1936–1938 and disappeared into the Soviet Gulag. Of these, some were either mobilized, or volunteered, to serve at the front during the war. At least a few are known to have survived the war and resumed a 'normal' Soviet life in the following years. The exiles who survived into the 1950s were allowed to return to their previous places of residence during Khrushchev's 'thaw', and received 'rehabilitation' from all charges against them. A handful were able to leave for Israel in the 1970s. Much of what we know about the lives of the Zionist exiles comes from these later immigrants and from the letters the exiles wrote to family, friends and comrades in Palestine.

The exiles' letters constitute a moving testimony to their stubborn effort to break the isolation imposed by exile and distance. They also document the heavy human toll exacted by the flourishing of Zionism in the 1920s and the depth of the exiles' commitment to their ideals. The hardships experienced by the exiles are often told in laconic language, but are harrowing all the same. In prison, the convicts suffered from cruel treatment, hunger strikes and suicides. In places of exile, they were repeatedly transferred to more remote hamlets, were subject to isolation, unemployment and hunger. They suffered debilitating illnesses, freezing cold in winter and scorching heat in summer. And most painful of all, many were condemned to watch helplessly the suffering of their children and loved ones. These deprivations became harder to bear as the years passed with no end in sight. During the Great Terror, in the second half of the 1930s, conditions worsened drastically for everyone caught in the Soviet network of prisons and labour camps.

The hope of leaving for Palestine remained alive among the exiles for a long time. Kuntsiia Tzap expressed this hope in a letter from Petropavlovsk, written in July 1934:

> A small group of comrades who received 'substitution' is leaving from Moscow any day now. Oh, the happy ones! My heart is happy for them. Will there be such happiness for me?! I still don't have an answer to this question, but hope is not lost.

She had been arrested in 1924 and again in 1928 for membership in Illegal Hechalutz and Hitachdut and had wandered among various places of exile and 'minus' zones. Denied 'substitution' in 1932 and again in 1933, she was rearrested together with other Zionists in Tashkent in 1937 and sent to prison camp, where the group was accused of organizing an uprising. Tzap was put in an isolation cell for two years, then sentenced to life-long exile in Krasnoiarsk, where she stayed until 1956. Finally 'rehabilitated' of all charges against her in 1959, she resumed her campaign for an exit permit and emigrated to Israel in 1969.[3]

When the hope for exit to Palestine was extinguished, the blow was almost unbearable. Zalman Sheptovitskii wrote from Karakala in December 1929, 'I informed you of the rejection of my request for "substitution". It affected me profoundly. Of all the disappointments I have experienced in my life, this last one is the most difficult and tragic.'[4] In March 1937, David Horenshtein wrote:

> After the tortures of waiting for an exit permit for 13 months I was turned down...If until now I had some hope, I now have none...I am already 38, and I have no desire to go to Palestine to die there...

Horenshtein was unusual among the exiles in that he had returned to Soviet Russia in 1927 after two years in Palestine, to support the work of Hechalutz. He was arrested within a year, sentenced to a three-year term in political prison, and his requests for substitution were denied at least three times (1933–1935). His fate remains unknown.[5]

While hope was still alive, the exiles exhorted their comrades in Palestine to do all they could to help their departure, 'Practical steps must be taken to obtain permission to leave for Palestine'.[6] In 1937, several members of Hitachdut still in Soviet Russia wrote to the movement's 'delegation' in Palestine about the conditions of several hundreds of Zionists in exile. All were incarcerated for 'membership in a counterrevolutionary organization' and expected to be rearrested immediately upon their release.

> Inform us in one way or another about the steps you are taking to save us, and whether there is any hope for a speedy release of the exiles...If you are unable to bring us out of here, our end will be bitter, for here, the concept of 'term of prison and exile' is highly flexible... *Under present conditions... we are condemned to a slow and tortuous death*. This description is in no way exaggerated...[7]

Occasionally, deliverance came late and was unexpected. The story of Chaya and Moshe Krimker (Asu'ach) is not typical, yet contains elements experienced by many other exiles. The two married in exile (Aulie Ata, Turkestan) and Moshe decided to forego his approved substitution and Immigration Certificate until Chaya completed her term of exile. In November 1929, the two moved to a 'minus' zone in Kursk and applied for substitution. Chaya, then pregnant, received an Immigration Certificate in May 1930, but Moshe's request to renew his earlier certificate was not approved. They decided that Chaya should leave all the same. Moshe was arrested again in August 1930 and exiled to Gorlan (Uzbekistan), then arrested there in August 1931 for sending money to a comrade, who used it to escape exile with the goal of returning to underground work. While in his third term of exile in Eniseisk, Moshe received a letter from Peshkova, who corresponded regularly with his wife, Chaya, in Palestine. Peshkova suggested that he apply to the State Political Directorate (GPU) in Moscow for substitution. In January 1934, the approval arrived, and in September, having secured the necessary papers and an Immigration Certificate, Moshe left for Palestine.[8]

For those who remained in exile, strong ties of fellowship were crucial to physical as well as spiritual survival. The Zionist exiles were drawn to each other, often married among themselves and otherwise lived in communal groups, united by comradeship and continued belief in Zionism. Eretz Yisrael was for many the focal point. Matilda Halperin wrote in July 1929 (in Hebrew) from the remote Priangarskii region:

> It is difficult for me to write to you who are in Palestine. When I write to exiles like me, I write mostly about the situation in Palestine [*ba-aretz*], about news received from there... My own life is so bland it's not worth writing about.[9]

The most common request in these letters was for more information about Palestine:

> Where does the present immigration come from? Its social make up? Are there new agricultural settlements and of what type, what share does Jewish labour hold in transportation, how do you train qualified workers, what is the state of schools and of physical education etc. etc.

The writer of these lines, Avnir Abrabanel, having reported in just one sentence about his daughter's tuberculosis, continues to list his questions:

> What are the dominant political groups and what do they demand of the Jews?... I read that oil refineries are to be built in Haifa, is it true? And what would be the share of Jewish workers? Is it true that a new railroad is being built from Haifa to Aqaba?[10]

The exiles expressed strong views on the issues discussed by the Jewish public in Palestine. Mendel Gervitz, for example, objected to the expansion of Jewish settlement to the eastern bank of the Jordan, advising instead to put all available resources into 'today's Eretz Yisrael'.[11] Avnir Abrabanel opined in October 1937 that 'the partition of Palestine is a lesser "evil"'. And Akiva Esterlis wrote about the Arab uprising of 1936–1939: 'One of the important conclusions we have to draw is the absolute need to attract to our side a certain segment of the Arab intelligentsia, and especially the Arab labourers. This important task belongs to our socialist public'.[12]

Ironically, it was in prison and exile that the Zionists refocused their attention from the struggle against Soviet dictatorship (an impossibly dangerous topic) to Palestine.

> I implore you to write to me from time to time about what is going on there, perhaps send a newspaper clipping. After all, it is easier to suffer when you know that, somewhere, a life dear to you is taking shape and friends who had not forgotten you are alive and well.[13]

Yakov Gorevoi who wrote these words was among the lucky ones – that same year, 1934, he was allowed to leave for Palestine. Others continued to depend on letters to satisfy their great thirst for information and they often complained about their interlocutors' silence. 'At long last I received news from the country that is so close to my heart, yet physically distant. To be honest, I waited so long for your letter that I despaired of receiving it.'[14] A letter from Gonia Vdovets (Vologda, March 1937) is even more reproachful, 'Your letter, like the preceding ones, does not satisfy me. I want to know much more.'[15]

Vdovets is among the few exiles reported to have eventually resumed a normal life. Arrested in 1929, he was denied substitution three times and came close to death while in exile in Archangelsk. He spent the 'minus' years in Vologda and was able to complete his university education, studying economics and sociology. Later, he headed the economic planning department of a regional medical centre, until retiring in 1968.[16] Another remarkable story of imprisonment and restoration belongs to Solomon (Sema) Liubarskii. A legendary leader of the left wing of Ha-Shomer ha-Tza'ir, he was released from prison in autumn 1924, after signing a declaration that he would cease Zionist activity. In early 1926, after an unsuccessful attempt to cross the border to Latvia, he escaped his guards by jumping off a train carrying him back to Moscow. Arrested once more in October 1926, he was sentenced to three years in political prison, followed by exile and restriction to 'minus' zones. Along the way, he completed his medical training and in 1941 volunteered to serve on the front, first as a combat soldier, then as military physician. After the war he settled in one of the large cities of Siberia, where he directed a hospital. Liubarskii died in 1978, never having seen the land for which he had worked and suffered.[17]

The letters from Zionist exiles in Soviet Russia, passed from hand to hand in the 1920s and 1930s, contributed to the symbolic image of Jews and especially Zionists in Soviet Russia as 'prisoners of Zion'. Theirs was a silenced and repressed Zionism, a far cry from their days of feverish activity in the 1920s and the practical, active Zionism that had been Russia's central contribution to Jewish nationalism in modern times. Is it possible that the ideas cherished in silence by the exiles and other hidden Zionists during the years of repression fed the awakening of Zionism in the Soviet Union after the Second World War and, again, in the late 1960's? To date, no one has explored this question.

If the Zionists in Soviet Russia were the 'prisoners of Zion', those leaving for Palestine became the practical as well as figurative 'builders of Zion'.

Their ideological, cultural and practical work in the Zionist youth movements and in the farms and workshops of Hechalutz had prepared them to embrace and claim this role. When they were in Russia, they were divided over the shape the new society should take and how best each of them could help its construction. Upon arrival in Palestine, many of them discovered how little they knew about the country, its inhabitants, the prevailing economic and political conditions and the public life of its Jewish population.[18] The immigrants responded to the reality around them in various ways, influenced by ties of comradeship as well as the political beliefs they had formed while still in Russia. Some built communal settlements, others cooperative villages and still others settled in cities and towns, where they played a major role in creating and staffing the organizations that framed the evolving society, especially the powerful network of organizations around the Histadrut.

While assessing the impact left by the substitution immigrants it should be recognized that they were not the only new arrivals from Soviet Russia during those years. Others included large families of Jews from Central Asia and the Caucasus, as well as urban Jews with savings, who could prove ownership of 400 or 500 British pounds per family member. Their capital entitled them to be admitted into Palestine under a separate category, not subject to the biannual allocations of Immigration Certificates. Chasidic followers of the Rabbi from Lubavitch also came, with help from their communities abroad and the Jewish Agency. During certain periods, the Soviet government was willing to allow such departures, largely for the same economic and political considerations that made the substitution emigration possible.

For our discussion here, it is especially important that the immigration from Soviet Russia during the decade 1924–1934 included many active Zionists – members of the Zionist youth movements and Hechalutz – who were indistinguishable from the substitution immigrants, except that their arrival did not entail time in prison or exile and did not go through the 'substitution' procedure. Some of them had crossed illegally into Romania, Poland, Latvia and made their way to Palestine from there. The Soviet authorities also allowed groups of expelled members of Tel Chai and other farms of Legal Hechalutz to leave for Palestine on boats from Odessa.[19] Counting these immigrants, roughly two and a half or three thousand active Zionists from Soviet Russia arrived in Palestine between 1924 and 1934.[20]

The lives of the immigrants followed several alternative paths, all of them intimately connected to the building of a Jewish society in Palestine.[21] Members of some movements made agricultural settlement their goal. Those from Illegal Hechalutz and the United All-Russian Organization of Zionist Youth or Histadrut (EVOSM) tended to join cooperative, non-communal settlements, such as Kfar Bilu or Kfar Oved,[22] though a group of latecomers from EVOSM did join the communal Kibbutz Mishmarot. Those who had belonged to socialist youth movements, if they chose agriculture, usually joined kibbutzim, as was the case with the leaders of Jewish Socialist Youth League (ESSM) who joined Ein Charod and Yagur. Members of the socialist wing of Ha-Shomer ha-Tza'ir initially formed a non-agricultural kibbutz, made up of several 'platoons' stationed wherever there was work, but unemployment during the economic decline of 1927 pushed them to accept agricultural settlement. The main group founded Kibbutz Afikim, whereas others joined Kfar Gil'adi and other kibbutzim.

Not everyone chose agriculture. In Ha-Shomer ha-Tza'ir, a movement with a high percentage of youth from big cities, the transition to agriculture caused many members to leave the kibbutz and settle in Haifa, Tel Aviv or Jerusalem. Likewise, members of the Zionist–Socialist party (TzS), which had done much of its work in Soviet Russia among the lower urban strata, chose to live and work in urban surroundings. The impact of these two groups was considerable. They found work in the interlocking economic, social and cultural organizations of the Histadrut: Tnuva, the marketing cooperative that became the single country-wide supplier of agricultural products from the cooperative and communal farms; Solel Boneh, the largest construction concern in the country; the Workers' Councils in cities and small towns and the services they operated for their members; Kupat Cholim, the health system of the Histadrut; Mo'etzet ha-Po'alot, the Council of Women Workers; Tarbut la-Am, the organization 'Culture for the People'; Hagana, the underground military organization; and of course, the organizational backbone of the labour parties, especially Achdut ha-Avoda and, later, MAPAI. The Zionist socialists found in these organizations an accommodating environment for implementing their Soviet-influenced understanding of how socialism could be built and, in turn, played an important role in shaping them.

A different course was followed by some of the immigrants who had obtained secondary or higher education in Russia. Especially among former members of the student movements (Hechaver, EVOSM, ESSM, Ha-Shomer ha-Tza'ir), a significant number continued their education in Europe, most of them in engineering, medicine, agriculture and other technical professions. Returning to Palestine, they usually worked in the public service sector, adding professional expertise to their earlier Zionist and social commitments.

It is difficult to summarize the lives of so many individual immigrants. Among those arriving as substitution immigrants were nationally known figures: the outstanding leaders of TzS and later Achdut ha-Avoda, Israel Bar-Yehuda (previously Idelson) and Zalman Aran (Aronovich); men who rose through the ranks to become ministers in Israel's governments or leading members of the Knesset, like Mordechai Namir (Nemirovskii) and Me'ir Argov (Grabovskii); the founders and leaders of the Women's Councils, Beba Idelson and Bat Sheva Chaikin; the popular composer Mordechai Ze'ira. Other men and women made names for themselves through work in their communities and professional circles. Examples of those include Olya Kaznachei, active in municipal affairs in Tel Aviv and at the Committee to Help Our Soldiers; her husband, Yehoshu'a (Shunya) Kaznachei, in his own words a 'simple worker' at the Tel Aviv branch of Tnuva, who was nevertheless chosen to represent Tnuva workers nationally, served on the Executive Committee of the Histadrut and was sent to Zionist Congresses;[23] Avraham Arest, who directed cultural affairs for the Jerusalem Council of Workers; Efrayim Tavori (Pipik), a sanitation worker in Tel Aviv, who held high posts in his party,

MAPAI, and was eventually elected to the Knesset;[24] Mania Pechenek-Levi who served as Chief of Nursing at Hadassah Hospital in Jerusalem.

Many, many hundreds of the Soviet Zionist immigrants of those years worked in relative anonymity alongside other newcomers, in agriculture, construction and industry. An overwhelming majority of the men served in the Hagana. In later years, a good number of them entered the growing Histadrut and state bureaucracies. Their private lives continued to be marked by their history in the Soviet Zionist movement. Most came to Palestine alone, without family. They married other members of the movement and maintained life-long friendships with those they had met as young people in the movement or in exile. Still, their thoughts and concerns focused on their new country and on the national life evolving there, and these concerns were all-consuming in the period leading to the establishment of the state and during its early years. They experienced themselves as full participants in the achievement of statehood and felt fortunate to have been able to exit Soviet Russia and reach Palestine. It was not only a matter of the natural inclination to justify retrospectively a crucial choice one had made early in life. In the late 1950s and early 1960s, during a relatively peaceful period in Israel's relations with the USSR, many of the immigrants visited their country of origin for the first time in three decades. Although their own living conditions were at best simple at the time, many visitors returned home shocked by their relatives' material circumstances and saddened by their isolation from Jewish life.

In relation to their past, these veterans of the Zionist movements in Soviet Russia acted differently as private people and as members of their old movements. In the private sphere, the veterans of Soviet Zionism behaved as if their commitment and success were sufficiently documented by the very foundation of a Jewish state and the social and political role played in this process by the parties and organizations they had helped build. They rarely told their children about their past in Russia – the years of war and pogroms, their youthful enthusiasm for Zionism, their underground work, arrests, exile and emigration – and only sparingly spoke of the difficult first years in Palestine. In behaving thus, they reflected the forward looking, positivist ethos of Israeli society and probably responded as well to their children's lack of interest. Collectively, however, each party, movement and organization active in Soviet Russia in the 1920s was determined to claim and document its singular contribution. Each published its own volume – a collection of memoirs or historical treatment.[25] The earliest works, published in the 1930s, are the most openly partisan, even contentious, whereas the latest, from 1981, takes a broader, more reflective view. All, however, are dedicated to staking a claim to their movements' responsibility for the Zionist revival of the 1920s. They express an underlying assumption about the extraordinary significance of the brief years of Zionist activity in Soviet Russia – the years that separated the enormous explosion in the influence of Zionism during the democratic revolution of 1917 from the repression and isolation of the last Zionists under Stalin's rule.

The movements of the 1920s appear in these volumes as both the culmination of Russia's practical brand of Zionism and as its increasingly endangered remnant.

The volumes revel in stories about Zionist underground work. They recall its many small victories over the Evsektsiia and in the cat-and-mouse game with the GPU. They pay tribute to the martyrdom of members who had died or otherwise lost the better part of their lives in prison and exile. What remains for the most part unspoken though strongly felt in these many pages is the sense of a body torn asunder. After all, only a relatively small number of Soviet Zionists reached Palestine and they found themselves cut off from their movements and from the Jewish population of the Soviet Union at large. In time, their sense of loss and bereavement (expressed with special clarity in the later, more reflective volumes) gave way to engagement in their new country, in a life of labour and public action and in the many struggles that ensued – against common, external enemies and within the national body politic. Their retrospective writings were as much a way of paying dues to the many active Zionists who were left behind, and of clearing consciences burdened by the guilt of survivors who had not been able to save their comrades, as they were a claim to the right of leadership in the new Jewish state.

Notes

1 Interviews by Ziva Galili with Fania Borovaia-Astrakhan and Badia Borovaia (January 1999); Fania Borovaia, 'Im "Ha-Shomer ha-Tza'ir" me-SSSR', Archive of Kibbutz Kfar Gil'adi, Astrakhan file. Some details regarding arrests, imprisonment and appeals for substitution of these and other individuals are based on information compiled by Boris Morozov from the archives of Pompolit.
2 Driker was mobilized in 1941 while in exile and died at the front. The daughter he had with Raia was raised by a relative in Moscow. Avraham Itai, *Korot Ha-Shomer ha-Tza'ir be-S.S.S.R.: No'ar Tzofi Chalutzi – NeTzaCh* (Jerusalem: Ha-Aguda le-Cheker Tfutzot Israel, 1981), p. 334.
3 Binyamin Vest, *Bein Ye'ush le-Tikva. Mikhtavim shel Asirei Tziyon be-Rusya ha-Sovyetit* (Tel Aviv: Reshafim, 1973), pp. 175–7; Out of his concern to avoid exposing comrades still living in the Soviet Union, Vest included in this volume mostly letters of those who had left the country or were known to have died.
4 Ibid., p. 230. Sheptovitskii had spent two years in political imprisonment (*politizoliator*) and was therefore not likely to be allowed a substitution of his exile.
5 Ibid., p. 66. Horenshtein had left Soviet territory in 1925 by crossing the Romanian border. When arrested, he was carrying false papers in the name of Chaim Baron or Baran. He signed his letters 'Chaim'. Ibid., p. 61.
6 The writer, Yechi'el Halperin (Yekhiel Gal'perin in the Russian documents), served repeated terms in prison and exile from 1927 to 1940, having been denied substitution, and continued to live in a 'minus' zone in Jambul, Kazakhstan, until finally allowed to leave for Kiev in 1955. He emigrated to Israel in 1970. Ibid., p. 89.
7 Lavon Archive, IV-104-53, f. 5. The folder contains many reports prepared by Hitachdut from 1924 through 1938.
8 Moshe Asu'ach (Krimker), 'Ha-Derekh le-Kfar Gil'adi' and 'Ani Moshik me-Tashkent', Archive of Kibbutz Kfar Gil'adi, Chaya and Moshe Asu'ach file; Itai, *Korot Ha-Shomer ha-Tza'ir be-S.S.S.R.*, pp. 286–7.
9 Halperin was arrested in 1928 for underground work in Illegal Hechalutz but received substitution in 1930. In Palestine she managed a training farm for women workers. Ibid., pp. 95–6.
10 Abrabanel was in a 'minus' zone in Kursk at the time (1935). Ibid., pp. 23–4.

11 Minusinsk, April 1935. Ibid., p. 79. Gervitz was arrested in 1930 for underground Zionist work and spent time in prison and exile. From there, he petitioned year after year for substitution, but it was repeatedly denied. His wife Bella who joined him in exile and habitation in a 'minus' zone lost all contact with him after his final arrest in December 1937. Ibid., p. 75.
12 Tas'evo, n.d. Ibid., p. 36. Esterlis was arrested in Moscow at the end of 1925 and spent many years in prison and exile, having been refused substitution. As late as 1948 he was said to be in a labour camp. Ibid., p. 32.
13 Tumen', 1934. Ibid., p. 72. Gorevoi (later, Gur-Avi) had been arrested in January 1931, while serving on the committee of the Labour Brigade of the Toiler's Wing of Ha-Shomer ha-Tza'ir. His earlier requests for substitution (1931, 1933) had been denied. Ibid., p. 71.
14 Akiva Esterlis, Tobol'sk, June 1929. Ibid., p. 33.
15 Ibid., p. 99.
16 Ibid., p. 98.
17 Itai, *Korot 'Ha-Shomer ha-Tza'ir' be-S.S.S.R.*, pp. 121–2, 225, 245. Liubarskii applied for substitution but was refused in 1927 and 1931.
18 This point was made in a report, prepared by the director of the residence for new immigrants (Beit Olim) operated by the PZE's Immigration Department in Tel Aviv, at the request of Colonel Kisch. The report (dated 9 January 1929) commented at length on the cultural and hygienic habits of the exiles arriving from Russia and their attitudes towards Zionism and Marxism. Central Zionist Archive (CZA) S25/2434/1–2.
19 Such groups left at the end of 1924, in winter 1925–1926 and in 1928 and 1929. Pines, *Hechalutz be-Kur ha-Mahapekha Korot Histadrut hechalutz be-Rusya* (Tel Aviv: Davar, 1938), p. 244; *Chalutzim Hayinu be-Rusya* (Tel Aviv: Am Oved, 1976), p. 465.
20 Of the nearly 1,300 members of the non-socialist labour movements in Soviet Russia (Illegal Hechalutz, EVOSM, Hitachdut) who arrived during that decade, less than 300 were 'exiles' (*golim*). This list does not include members of the socialist movements, including Legal Hechalutz. Lavon Archive, IV-104-53, f. 67.
21 This 'portrait' of the pioneer immigrants from Soviet Russia is based on interviews by Ziva Galili with Shlomo Ben Shoshan, Nelli Bliumkina, Badia and Fania Borovaia, Klara Galili, Lyova (Arye) Golani, Olya Kaznachei, Shoshana Levin, Syoma (Shlomo) Pinskii and Pnina Yeiny. Also consulted were many interviews at the Oral Documentation Department, Institute for the Study of Contemporary Jewry, Hebrew University, Jerusalem (projects 42, 129 and 131) and many published and unpublished memoirs.
22 I am grateful to Ami If'ar for this information.
23 When he died in 1960, Kaznachei's funeral procession travelled among three centres of power in Tel Aviv of those days – the organizational headquarters of MAPAI (Beit Arlozorov), the Histadrut and Tnuva. At each stop he was eulogized by the heads and leaders of these organizations. Yehoshu'a (Shunya) Kaznachei, *Alav u-Mishelo* (Tel Aviv: Ha-Irgun ha-Artzi shel Ovdei Tnuva, 1963).
24 To memorialize him, the cultural centre of the Tel Aviv Workers' Council was named after him, Beit Tavori. Efrayim Tavori, *Mi-Sha'ar el Sha'ar. Divrei Hagut ba-Mishna ha-Tziyonit-Sotzialistit* (Tel Aviv: Mo'etzet Po'alei Tel Aviv-Yaffo, 1959).
25 For a list of these, see the Introduction to this volume, note 10.

5 Documents

1 Mandate given to E.P. Peshkova by the Deputy Chairman of the GPU

GARF f. 8409, op. 1, d. 11, l. 1
[Moscow], 11 November 1922

1   E.P. Peshkova is hereby granted the right to render assistance to political prisoners and their families.

2   The assistance mentioned in paragraph 1 is to consist of:
    (a) material assistance to political prisoners and their families, in the form of items, foodstuffs and funds within the limitations of prison regulations
    (b) initiation and pursuit of a process of appeal to the GPU on matters concerning prisoners and their families

3   In pursing these goals, Peshkova is authorized:
    (a) to appeal personally to the empowered representative [*upolnomochnyi predstavitel'*] of the GPU with the oral and written petitions mentioned in point 2 and to receive responses to these appeals
    (b) to send to places of incarceration mentioned in point 2, parcels and donations to the prisoners from her as well as from relatives of political prisoners

4   E. Peshkova is authorized:
    (a) To receive from political prisoners via the responsible prison authorities and in connection with those still undergoing a process of investigation – via the investigative bodies of the GPU, in accordance with point 3a above, information concerning those who require assistance
    (b) To receive appeals addressed to her, including by mail, from relatives of political prisoners
    (c) To organize concerts, performances, collect funds, receive donations, purchase, sell or exchange in order to raise funds for these activities, within the boundaries of existing laws and regulations.

(d) To maintain a professional staff to deliver aid, office space, storage space, stationery and a stamp with the imprint: E.P. Peshkova, Aid to Political Prisoners

5   In the framework of the above, the organs of the GPU are to offer E.P. Peshkova cooperation by reviewing her petitions as matters of first priority.

[Signed]
I.S. Unshlikht[1]
Deputy Chairman of the GPU

■ ■ ■ ■

2   Memorandum of A. Merezhin, Central Bureau of the Evsektsiia to the Politburo

RGASPI f. 445, op. 1, d. 119, ll. 120–1
[Moscow, March 1923]

To the Politburo of the CC RCP(b)

In connection with the response received from the CC CPU (attached here)[2] to the request of the CC RCP concerning Hechalutz, at a session of the Orgburo of 19/3, the question of Hechalutz was raised once again and the Orgburo, in place of its decree of 30/8 [1922] concerning the liquidation of Hechalutz, has ruled

To authorize the existence of the organization Hechalutz in the general framework established by the Soviet regime, while granting the GPU the right to fight counterrevolutionary elements which are present in Hechalutz.

Considering this decision to be a great political mistake, the C[entral] B[ureau] of the *evsektsii* of the CC RCP, requests that the Politburo bear in mind, apart from the considerations expressed in previous memoranda concerning Hechalutz, which were forwarded to the Orgburo and the Secretariat, the following:

1   The latest ruling of the Orgburo is tantamount to legalization of the entire Jewish bourgeois public and of the strongest Jewish Zionist bourgeois party.

2   This legalization affects 90 per cent of the Ukraine, Belorussia, the Western *guberniias*[3] of Vitebsk, Gomel' and Smolensk.

3   In these *guberniia*s, where either a majority or close to a majority of workers in Soviet enterprises are Jewish, the overwhelming majority of merchants and speculators are Jews as well. In these circumstances special caution is required on the part of the party and the Soviet organs, since workers, Red Army and peasant masses do not identify Soviet power with Jewish power and have not been infected with hatred for the Soviet organs nor with antisemitism. Any overtly favourable approach to the Jewish bourgeoisie threatens discrediting the party in the eyes of labouring non-Jews, no less, and even more than Jews.

4   Therefore, the question of legalizing Jewish petty-bourgeois and nationalist organizations is closely linked with questions of legalizing non Jewish organizations

as well, in particular Ukrainian and Belorussian nationalist organizations of a Petliurian and Balakhovite bent.[4] But insofar as this question is decided negatively (and in border *guberniia*s it cannot be decided otherwise) the former question must be decided negatively [as well]. Any other resolution of these questions *places our local organs in an impossible situation*.

5   Hechalutz is striving also to create a system of Jewish national agricultural, producers' and credit cooperatives. By doing so, national cooperatives of Russians, Ukrainians and Belorussians will unavoidably emerge and will split the cooperative movement along national lines and bring about national squabbles and, in particular in the Ukrainian circumstances, will demoralize everyone and everything.

6   The view that the Jewish bourgeoisie, in and of itself, does not pose a danger to Soviet power is true. But the problem is that the Jews organize not only themselves, but others. The Bund is the most organized, strongest faction of the Russian Social Democratic Labour Party (Mensheviks) and [this faction] is responsible for organizing [the RSDLP] to a much greater degree than non-Jews. And the Zionists, who have been legalized through Hechalutz, being a force of national discord, could, at the same time, become consolidators of non-Jews, quasi-SRs, quasi-Kadets and quasi-*Smena Vekh*[5] elements (with whom they come into contact in Soviet institutions, in trusts and in commercial operations) for the struggle against Soviet regime – not by means of armed uprisings, but through its economic and ideological disintegration. (e.g. the speech of Prof. Brutskus at last year's agronomy congress.[6] Prof. Brutskus is a nationalist in Jewish questions but he allies himself with non-Jewish forces for the political and ideological onslaught against Soviet power). In this way the Hechalutz organizations undoubtedly would become connecting and organizing centres not only for Zionist elements, but for various non-Jewish, anti-Soviet elements as well.

It should be added that, at present, in central Russia, on the Volga, in the Urals, in Siberia and above all in Moscow and Petersburg, there are many more Jews than there were before the war, and that they have a relatively significant number of Zionist organizations (if not in numbers of members, then in the number of centres) including those outside the former Pale of Settlement.

7   There are those who say that a more favourable relationship towards the Jewish bourgeoisie in Russia attracts Jewish foreign capital to Russia in the form of loans, etc. The Central Bureau considers this view to be completely fallacious. If foreign Jewish capitalists do not currently invest their capital in Russia, it is true for the same reasons that capitalists of all other nations do not currently invest their capital in Russia – it is because of the Soviet regime. The depiction of the Jewish bourgeoisie as some kind of special bourgeoisie, which is ruled according to national considerations has no basis in reality. Jewish foreign bankers (Rothschilds, Mendelssohns) loaned money to the autocracy in spite of the highly unjust circumstances of the Jews which prevailed in Russia at that time, in spite of the Kishinev pogrom, and in spite of and immediately in the wake of the horrible pogroms of 1905. [The Jewish bourgeoisie] sometimes protested against the overt and increasing scale of the pogroms, but immediately following these protests, as soon as the pogroms turned from 'overt' to 'hushed', they continued to finance even the Russian autocracy and the Romanian government and others. Only recently the

Jewish Colonization Association (JCA),[7] situated of late in Paris, gave a subsidy to Wrangel for moving his troops from Constantinople, in spite of the fact that Wrangel spilled no small quantity of Jewish blood.[8] It is characteristic that Jewish foreign bankers, who are favourable to Zionism, feed the Zionist movement with small gifts and donations, but do not invest their capital in Palestine, for they consider it unprofitable. As a result of certain historical conditions, the Jewish bourgeoisie is generally the most cowardly and most 'patriotic' bourgeoisie (in relations with the bourgeois governments of their own countries) and would never do anything that might arouse dissatisfaction on the part of the non-Jewish bourgeoisie surrounding it or which would make it suspect of having Bolshevik sympathies.

Thus, the legalization of the Jewish petty bourgeois groupings in Russia creates complications in Russia, and does not offer any benefits in terms of attracting foreign capital.

8 Based on the foreign Jewish press, one must conclude that there are no grounds for any hope of exploiting the sympathies of the foreign Jewish petty bourgeois masses, since there has been an internal rift there lately and their mood can be characterized approximately as follows: 'we must reject the forcible overthrow of Soviet power because pogroms are an unavoidable concomitant of this, but everything must be done in order to break it from within, by economic pressure on the one hand, and by creating in Russia a variety of organizations (disguised as charitable), which serve as a fulcrum for Jewish and non-Jewish collaborationist *Smena Vekh* elements, on the other. This explains to a significant degree, the influx of the various foreign philanthropic organizations into Russia.

9 A benevolent attitude towards the Jewish bourgeoisie will also incite increasing nationalism even among workers who are not fully conscious, but are less revolutionarily-oriented, and who will interpret this as some kind of deprivation (depriving them of the weapons of struggle, which non-Jewish workers use) and as a justification of Jewish nationalism as progressive.

Thus, the Central Bureau of the *evsektsii* of the CC, RCP requests that the Politburo repeal the decision of the Orgburo of 19/III and retain the Orgburo decision of 29/VIII which resolves: 'To concur with the resolution of the Narkomindel[9] and the GPU regarding this question. Convey the current decree to the Presidium of VTsIK[10] for implementation.' In the wake of this decree, the following coded telegram was distributed (4/IX)[11]:

To all Gubkom[12] secretaries

In view of the ongoing liquidation of the Jewish Zionist organization Hechalutz, it is necessary to take control of the *zemkhozy*[13] and collectives created by it, and insert Jewish worker-communists and reliable non-party members there. Work in close contact with the *evsektsii*.

[signed]
Secretary of the C[entral] B[ureau] of the Evsektsiia, CC RCP
Merezhin

■ ■ ■ ■

3 Report of the OGPU on the search and arrest of Zionist Activists

RGASPI f. 17, op. 84, d. 643, ll. 4–6.
[March 1924]
Top secret

On the night of the 13–14 of this month, the secret section of the OGPU carried out searches of active members of the Zionist organizations: Algemeyn Tziyen[14] [General Zionists]; TzS TzTz (Zionist–Socialist Party Tze'irei Tziyon); STP TzTz (Zionist Toilers' Party Tze'irei Tziyon); Hechaver (the student Zionist organization), which brought quite significant results in that the materials discovered completely confirmed the intensified activity of these anti-Soviet illegal organizations in the recent period.

1 In almost all the searches illegal Zionist literature was found: newspapers, journals, bulletins, circulars and other party materials, both from abroad and locally produced.

2 At the apartment of the secretary of the Moscow committee of TzS TzTz an illegal meeting took place. At the moment when our commissar appeared, the members of the Central Committee and other active workers of this organization on the scene rushed to tear up into small pieces some kind of papers which are now being restored. Evidently these were illegal Zionist materials.

Under the table was found a briefcase which contained handwritten protocols of an All-Russian Congress which took place illegally in Moscow at the beginning of this year.[15] Everyone denied having anything to do with the briefcase. At the same meeting, the Zionist Gelbets was arrested. He was supposed to have been exiled and had hidden from the Crimea GPU.

3 At the residence of Shub,[16] a member of the Central Committee of Hechaver (the student organization), illegal bulletins were found in the Jewish language, reproduced on a duplicating machine in a large quantity (over 80 copies of the latest issue). The latest issue was dated 17 March – prepared, apparently, to be distributed by that date. This bulletin belonged to the newly reorganized illegal Zionist Socialist Youth League, Tze'irei Tziyon,[17] which has been operating intensively from the first day of its existence. The most recent issue of *Sotsialisticheskii vestnik* was also found in his possession.[18]

Note: the TzS TzTz reprints material from *Sotsialisticheskii vestnik* in its own illegal publications.

4 These same bulletins were found in the possession of a number of other individuals.

5 Miss Malkova,[19] an active member of the Moscow Committee of TzS TzTz was found in possession of a large amount of illegal Zionist materials and a typewriter with a chemical ribbon.

6 Mr. Shpiro[20] was found in possession of a mimeographed illegal circular of the Zionist Socialist Youth League Tze'irei Tziyon.

7 A member of the Central Committee of the illegal TzS TzTz, Mr. Aronov,[21] who is at the same time a member of the Central Committee of the currently legal Hechalutz, and who resides at the office of the Central Committee of Hechalutz on Lukovoi Lane, was found in possession of a duplicating machine, which, according of the explanation of another member of Hechalutz, Mr. Pines,[22] is supposedly in disrepair and can not be used. This, however, is not consistent with the facts since, according to the data of our agency, the Central Committee of Hechalutz was putting out illegal bulletins printed on a duplicating machine. The latter, incidentally, was freshly inked.

This operation confirmed our agency's information that the Zionist organizations, especially the most anti-Soviet among them, the TzS TzTz and STP TzTz, have, in the recent period, been operating energetically, and after the August crackdown in Kiev (eight duplicating machines and one mimeograph machine were confiscated and the most active members were exiled), they transferred the centre of the illegal activity to Moscow, developing it particularly strongly among young people and students.

Forty nine people were arrested in the operation. Of them, the most active and high ranking were:

1 Sh.G. Aronov – member of the Central Committee of TzS TzTz, member of the Central Committee of Hechalutz, unemployed, intelligentsia.[23]
2 L.I. Levkin – member of the Central Committee of TzS TzTz, an engineer at Gosposrednik,[24] intelligentsia.
3 N.I. Plotkin[25] – secretary in charge of the TzS TzTz Moscow Committee, accountant at the publishing house Krasnaia Nov',[26] intelligentsia.
4 Ia.M. Iavno[27] – member of the Central Committee of Hechaver, student at the Agricultural Academy, intelligentsia.
5 L.Z. Shub – a leader of the Zionist Socialist Youth League, a student at the agricultural academy, intelligentsia.
6 Gurvich[28] – a leader of the Zionist Socialist Youth League, a student at the agricultural academy, intelligentsia.
7 I.P. Mendel'son[29] – member of the Central Committee of the Zionist Toilers' Party and chairman of Right Hechalutz, unemployed engineer, intelligentsia.
8 B.G. Vest [West][30] – member of the Central Committee of the STP TzTz, unemployed, intelligentsia.
9 I.L. Gol'dberg[31] – member of the TzK [CC] Algemeyn Tziyen; member of the board of the Jewish Society, intelligentsia.
10 E.A. Belkovskii[32] – member of the Central Committee of Algemeyn Tziyen, legal advisor to TsULP,[33] intelligentsia.
11 L.S. Lebedinskii – member of the main bureau of TzS TzTz, responsible secretary for the Commission for Aid to Children of the Moscow City Council (Moscow Soviet), intelligentsia.
12 R.P. Gel'chinskaia[34] – member of the Moscow Committee of TzS TzTz, unemployed, intelligentsia.

13 S.B. Malkova – member of the Moscow Committee of TzS TzTz, student, intelligentsia.
14 I.Z. Liubetkin – member of the city bureau of TzS TzTz, statistician at the MSNKh.[35]
15 D.A. Chudakov – Former aide to the General Secretary of Algemeyn Tziyen, student, intelligentsia.

The others are also active members of the above-mentioned illegal organizations. Attached: samples of the confiscated illegal literature.

Assistant Chief of the OGPU
[signature] Iagoda[36]

Head of the OGPU Secret Department
[signature] Deribas[37]

■■■■

4 F.E. Dzerzhinskii to his Deputies V.R. Menzhinskii and G.G. Iagoda

RGASPI f. 76, op. 3, d. 326, ll. 2–3
15 March 1924

Comrades Menzhinskii[38] and Iagoda,

I reviewed the Zionist materials. I must admit, I do not understand at all why they are being persecuted on the basis of their Zionist affiliation. The majority of their attacks on us are based on our persecution of them. Persecuted, they are a thousand times more dangerous for us then than they would be not persecuted, developing their Zionist activity among the Jewish petty and large scale speculative bourgeoisie and intelligentsia. Their party work, therefore is not at all dangerous for us. The workers (the real ones) will not follow them, but their cries connected with their arrests will reach the bankers and 'Jews' of all countries, and will do us no small amount of harm.

The program of the Zionists is not dangerous to us; on the contrary, I feel it is useful. I was once an assimilator.[39] But this is a 'childhood illness.'[40] We should assimilate only the most insignificant per cent; this is enough. The rest should be Zionists. And we should not bother them, on the condition that they do not interfere with our policy. Let them be allowed to criticize the Evsektsiia and the same goes for the Evsektsiia. But then we must mercilessly beat and punish the speculators (scum) and all those who violate our law. We will reach out in a similar way to the Zionists[41] and try to give positions not to them but to those who consider the USSR and not Palestine to be their homeland.

F[eliks] D[zerzhinskii]

■■■■

5 Y. Mereminsky, Immigration Centre to Central Office of World Hechalutz (Berlin)

Lavon Archive IV-211-3, f. 2
[22 October 1924]

Dear Comrades,

We hereby acknowledge receipt of your letter from the 15th of this month, number 1052, together with a list of comrades who are under arrest in Russia [...]

The issue of the imprisoned comrades in Russia is of continuous concern. Every day we receive telegraphed demands from Hechalutz (Oscher), Ha-Shomer ha-Tza'ir, and from the various political movements asking that we send them, by telegraph, entrance permits. Much to our chagrin, we are limited in what we can do since permits that were issued by the government several months ago were used in bad faith.

The Zionist Executive would, in the past, receive permits for political prisoners beyond the 'schedule' of certificates.[42] Numerous permits were sent by telegraph as well. Suddenly it was discovered that these permits were being used by those who had no connection whatsoever to the political prisoners. Someone had simply telegraphed the order to Palestine to 'issue permits for the prisoners', and then had received them.

All of this has created a situation in which the government does not issue a single permit by telegraph. In addition, the government places all responsibility on the shoulders of the Zionist Executive and requires that it subtract these certificates from those it *will* be receiving over the next six months.

The situation has become so difficult that we are receiving no permits at all, neither for prisoners nor even for comrades who have already been tried and are awaiting imminent – 24 October, 30 November – deportation to Siberia. They are not even issuing entry permits to those comrades under arrest in Russia who have siblings here.

We are obviously doing everything possible in order to solve this crisis. Our hope is that we will receive permits for 22 comrades (Nemirovskii[43] and others) who are to be deported the day after tomorrow.

We promise the Zionist Executive to be vigilant and employ all means available in order to ensure that permits intended for political deportees will be given only to them. Colonel Kisch[44] promised us, in reply to our demands, to reacquire from the government in Palestine a *limited* number [of permits] for the prisoners whose sentences have already been pronounced in Russia and who are soon to be sent to Siberia. The results are not yet known to us.

We presently have a list of *60 to 70 persons* who have been arrested. In addition, we are adding to the list the 9 comrades about whom you have written to us.

If we do not receive any permits beyond the norm, then all 70 persons will be the first candidates to receive the certificates we hope will soon be arriving. (300 to 400 for Hechalutz[45] over the next six months). If you keep in mind that both we and the Zionist Executive receive hundreds of requests each day from Odessa,

Kharkov, Minsk, and Moscow to issue visas for the *prisoners*, then you will understand how difficult it is.

We are also writing today to [Hechalutz in] Russia and giving them detailed instructions.

We will inform you separately when they send the permits (by post, not by telegraph) for the 9 comrades listed above.

At the same time, we must point out to you that for a month now all immigration affairs have been concentrated in the Immigration Centre, and you must demand that all the central committees of Hechalutz [in different countries] contact us immediately, and that they also exert similar influence on the labour departments attached to the Palestine Offices.[46]

We are in direct contact with Russia. Nevertheless, if you receive news or requests related to immigration matters we ask that you send them on to us and together we will make a concerted effort to help our comrades there.

[...]

With comradely greetings,

Y. Mereminsky

■ ■ ■ ■

6 Y. Mereminsky, Immigration Centre to Palestine Zionist Executive

Lavon Archive IV-211-3, f. 17A
10 December 1924

Immigration Department[47]
Strictly private
Urgent

Executive, Zionist Organization
Jerusalem

Sir,

In accordance with the conversation which our Com. Y. Mereminsky had with Prof. Pick and Mr. Jacobs, we wish to submit to you herewith some information on the conditions in Russia, on our comrades of the enclosed list and a citation from their letter.

A proclamation issued by the Zionist Socialist Party, containing a protest against persecutions of Zionism in Russia, caused the arrest of 3000 persons, of them 400–500 are still imprisoned.[48]

Considering that the number of the certificates is limited, the Zionist Socialist Party do[es] not want to ask for such before the persons are tried and their fate is known.

The enclosed list contains only names of persons tried already: Dr. Bathia Loulka – 3 years banishment to Solovetsky Monastery, Tania Fireman – 2 years of Concentration Camp, Barou-Hazanovitch – deportation from Russia for 3 years, Weisman Butsikowsky and brother – the same, Anatoly Grinfield. Gutta Bassevitch and others – 2–3 years Siberia, etc.

Our comrades in the prisons of Harkow [Kharkov], Moskow, Winiza [Vinnitsa], Odessa, etc. protested against these sentences by means of hunger strike and they were cancelled. A telegram dated 6.12.24 informed us that the comrades Promislov, Budnik, Kranz etc. (11 persons) will be sent to the places of their destination on the 16.12.24. A letter dated 18.11 informs of 23 persons (Dr. Bathia Loulka etc.) and last letter tells of comrades Sverdlov and Rosenstein, who have been our immigration agents and now must leave immediately Russia.

The following citation from the letter dated 14.11 will give an idea of the conditions of the above mentioned hunger strike:

> On 1.11 the prisoners demanded categorically the investigation shall be terminated, otherwise they go on strike. The answer was that they were not allowed to use the W.C. The prisoners made a noise and all other prisoners struck too. Soldiers were brought in, and four (among them Bathia Loulka) were dragged away and isolated. Only on the third day the Procurer[49] came and let them know the sentences. Four persons, who were sick with anguine,[50] were not admitted to hospital, etc.

If these comrades would not receive immediately certificates they would be deported on 16–17 this month (which means 5–6 months of winter journey from prison to prison) and then the certificates, even if sent, would not reach them before half a year.

We ask for extraordinary arrangement of this matter.

We have to add: To all the girls who are on the List certificates have been sent by post a week ago. On the same day certificates were sent to Grinfeld Anatoly and Promislov, Jacob, but the papers could not reach them in time.[51]

Yours respectfully,

Y. Mereminsky

■ ■ ■ ■

7 F.E. Dzerzhinskii to V.R. Menzhinskii

RGASPI f. 76, op. 3, d. 326, l. 4
24 March 1925

To Comrade Menzhinskii

Is it right that we are persecuting Zionists? I think that this is a political mistake. Jewish Mensheviks, i.e. those working among the Jews are not a danger to us. On the contrary, this is not an advertisement for Menshevism. We need to reconsider our tactics. They are incorrect.

F[eliks] D[zerzhinskii]

■ ■ ■ ■

8 Report of Special Department of OGPU to F.E. Dzerzhinskii

RGASPI f. 76, op. 3, d. 326, l. 5
Top secret
29 May 1925

To Comrade Dzerzhinskii

1 Throughout the entire USSR, according to the data available at the present moment, 34 Zionists are under arrest (In Moscow, 1; Minsk, 32; Rostov, 1).

2 132 people have been exiled within the USSR. Of them, 8 people have been sent for 2 years and 124 people, for 3 years. They have been exiled primarily to the Kir[giz]krai, Siberia and the Urals.

3 A total of 13 people are incarcerated in a concentration camp for a period of three years each.

4 A total of 152 people have been exiled abroad and permitted exit as a substitution for exile. With regard to the question of denying permission to substitute [internal] exile with departure for Palestine, we adhere to the following tactic: the most active elements, members of the Central Committee and the Provincial Committees, who have been found to be in possession of serious items such as anti-Soviet leaflets, public appeals and printing presses, are not to be permitted to leave for Palestine. The less active elements will be permitted to leave for Palestine. This tactic is based on past experience in the struggle against the Zionists.

When, up to the end of 1924 we were primarily deporting [Zionists] to Palestine, this became a major stimulus for intensifying national activities of the Zionists, since each one was sure that for his anti-Soviet activity he would receive the opportunity to travel at public expense (of the Zionists or organizations sympathetic to them) to Palestine without paying the price for the crime he had committed.

We impose exile and concentration camp primarily on the active members of TzS (the Zionist–Socialist Party) which in essence is a Menshevik party.

Chief of the Secret Department, OGPU
Deribas

Chief of the 4th Section of the Secret Department, OGPU
Genkin[52]

[On the same folio the following notation appears:]
31 May 1925

To Comrade Menzhinskii:

All the same, I think that such a broad persecution of Zionists, particularly in the border regions will not bring us any benefits either in Poland or in America. It seems to me we need to influence the Zionists so that they renounce their counter-revolutionary operations with regard to Soviet power. After all, in principle we could be friends with the Zionists. This question has to be studied and raised in the Politburo. Zionists have a large influence both in Poland and in America. Why have them as our enemies?

F[eliks] D[zerzhinskii]

■ ■ ■ ■

9   A.M. Hyamson, Chief Immigration Officer, Government of Palestine to British Chargé d'Affaires (Moscow)

ISA 11/1147, CONS/E/37/2
16 September 1925

The British Chargé d'Affaires, Moscow

Sir,

I have the honour to state that within the last two or three months a number of applications have been made for the authorization of visas for Palestine as a matter of great urgency to persons in Russia who are stated to be prominent Zionists in great danger at the hands of their Government on account of their Zionist activities. These applications have been made not on behalf of single individuals but of groups of thirty or forty individuals or families at a time. There is of course no objection to the grant of visas to Palestine to prominent Zionist who qualify under the Palestine Immigration Regulations, provided that you do not consider them politically or medically undesirable so far as Palestine is concerned, but requests have been and are being urged that all conditions should be waived and in view of the urgency of the cases authority for the visas should be telegraphed to you forthwith.

I shall be much obliged if you would be so good as to let me know whether there is at present or has been recently any special persecution of individuals on account of their Zionist sympathies and whether the statement that certain individuals were and are in great danger on account of those sympathies is correct.

I shall be much obliged for an early reply.

I have the honour to be,
Sir,
Your obedient servant.

Sgd. A.M. Hyamson[53]
Controller

■ ■ ■ ■

10   Central Eretz Yisrael Office (Warsaw) to Immigration Department, Palestine Zionist Executive (Jerusalem)

Copy: CZA S25/f. 2424
Warsaw, 10 Nissan 5687, 12 April 1927

Immigration and Travel Section

Dear Sir,

On her way from Moscow to Rome, the president of the Political Red Cross in Russia, Mrs. Ekaterina Gorky Peshkova, stopped in Warsaw.

She invited comrade Y[israel] Ritov to meet with her there and discuss the condition of the Zionist prisoners in Russia. She informed us that no less than one hundred certificates are required immediately for the prisoners – members of Hechalutz, TzS, etc.

She also informed us that the material conditions of the Zionist prisoners is no less than catastrophic. Financial assistance is not to be had from anywhere, and the prisoners have access only to the meagre sums received by the Red Cross from donations originating with the SRs and SDs. The situation has caused frustration among these parties since they bear the burden of caring for their own imprisoned comrades. They also claim that it is neither just nor honest for the World Zionist movement to place the responsibility for Zionist victims in Russia on the shoulders of other political parties that lack in resources.

Mrs. Peshkova will be visiting Warsaw again on her way back to Moscow (in several more weeks) and she asked that she be then given a concrete answer.

We are bringing this to your attention and are awaiting your response, to be delivered to Mrs. Peshkova.

Respectfully,

Shefer and Ritov [signed]

P.S. Mrs. Peshkova asked that the above conversation not be made public.[54]

■ ■ ■ ■

11 A.M. Hyamson, Chief Immigration Officer, Government of Palestine to F.H. Kisch, Chairman of the PZE

ISA 11/1174, IMM/7 I
Jerusalem, 30 June 1927

Lieutenant-Colonel F.K. Kisch, D.S.O., O.B.I.
Palestine Zionist Executive

Dear Kisch,

I have had your letter K/2455/27 of the 22nd instant regarding the period of validity of Immigration Certificates sent to Russia.

The new facts since my communication of the 11th of March 1926 are the following. In the first place a new Immigration Ordinance which lays down that visas authorised under the Schedule must be granted not later than the 31st of March and the 30th of September respectively, has come into force. The earlier instruction to the British Chargé d'Affaires at Moscow referred to in my communication of the 11th of March is not specifically cancelled and I believe he continued to act on it, but as you will realise it is not possible to instruct him to

98 Documents

ignore the instructions that have been issued by the Foreign Office and should bind him. Moreover, since March 1926 the practice of placing a certain number of visas at the disposal of the Political Red Cross in Moscow has grown up. As a consequence they have always had some at their disposal. That is to say some presumably lapsed on the 31st of March last, but a new supply was available on the 1st of April. Therefore, even if the law is observed the Political Red Cross has not I think for the past year ever been at a loss for an Immigration Certificate urgently required for a candidate who can fulfil the conditions laid down.

Unfortunately however the whole discussion is now academic. You are I believe aware that no visas for Palestine can now be granted in Russia.[55] Those at the disposal of the Political Red Cross are therefore useless and it should be well to recover from them the Immigration Certificates that have not been used if you can do so. We are endeavouring to make some arrangement whereby residents of Russia can obtain their visas in another country, but as you can realise the difficulties are considerable. If you can suggest a British Consulate or Consulates which residents of Russia can reach with a minimum of difficulty, I shall be grateful to you.

Yours sincerely,

Sgd. A.M. Hyamson

■ ■ ■ ■

12   Memorandum of Norwegian Legation on the procedure for obtaining Palestine visas for Soviet citizens

ISA 11/1174, IMM/7 I
London, 18 July 1927

Norwegian Legation
London.
1 enclosure.

The Norwegian Minister presents his compliments to the Secretary of State for Foreign Affairs, and, acting under instructions from his Government, has the honour to transmit herewith a *promemoria*[56] from the Norwegian Legation in Moscow regarding the procedure for obtaining Palestine-visas for Soviet subjects.

His Majesty's Minister in Moscow states that he shall feel greatly obliged for receiving the answer of the British Foreign Office to the question raised in the *promemoria* at an as early date as possible.

Promemoria

Regarding the procedure of obtaining Palestine-visas for Soviet subjects.

A representative of the organization 'Assistance for Political Prisoners – prev. Political Red Cross – E. Peshkova' has called at the Royal Norwegian Legation at Moscow presenting a dozen Soviet passports for prisoners of Jewish faith at present committed to prison in connection with political matters and requesting the assistance of this legation in order to obtain for them collective visas for Palestine in the same manner as previously arranged by the British Mission in Moscow.

The representative was informed that according to temporary instructions the procedure for obtaining single visas was as follows:

Applicant receives from this legation necessary forms which he fills out and which he himself sends to The Chief Immigration Officer, Jerusalem. Applicant at the same time declares what route he will follow. He will then receive direct information whether he gets the requested permission or not and what British station he will have to get in touch with in order to obtain the necessary visa. He then writes directly to the station indicated and from it receives instructions how to proceed to obtain the necessary first visa (for all political purposes this legation presumes there can only be the question of Constantinople). On presenting himself before the competent British official indicated the applicant will then receive his Palestine visa.

The Royal Norwegian Legation further informed the representative of the organisation mentioned that it had no instructions whatever regarding the matter of collective visas for Jewish political prisoners who would be released by the Soviet authorities if they can obtain a visa for Palestine, but that the legation would take measures to obtain instructions regarding this matter. The representative of the organisation most earnestly requested that such instructions be obtained as quickly as only possible on account of this being a matter of vital importance to many people.

The Royal Norwegian Legation in connection with the matter of Palestine visas would consider it of importance to obtain instructions regarding the following points:

1 Can persons who have already received information from a competent authority that their applications have been granted write directly to the British Consulate General at Constantinople for all further information and
2 Will the letter from the Chief Immigration Officer at Jerusalem or from the British Consulate General at Constantinople be sufficient documentary evidence to procure for them the necessary Turkish visa in order to proceed to Constantinople?
3 What will the procedure be in case the British Authorities decide to grant collective visas for Jewish political prisoners?

Moscow July 7th 1927

■■■■

13 F.H. Kisch, Chairman of PZE to M. Nurock, Assistant Chief Secretary of the Government of Palestine

CZA S25/f. 2424
26 July 1927

Dear Nurock,[57]

With reference to your telephone call of yesterday, as I informed you over the phone Madame Peshkoff [Peshkova] is President of the Political Red Cross at

Moscow, a body which has been in existence now a considerable time, having been established in order to try to protect victims of Soviet political persecution. Madame Peshkoff is the wife of Maxim Gorky, and is a personality who not only commands such respect from the Soviet Authorities that the latter do not interfere with her work, but has secured a measure of recognition for her institution from the principal European Governments.

From time to time over the prolonged period we have saved Russian Zionists from exile to Siberia through the timely intervention of the Political Red Cross, and Hyamson can give you further information on the subject.

With regards to Madame Peshkoff's present request to the Norwegian Minister for visas for 81 persons,[58] I cannot support the despatch of such a number of visas without knowing precisely who they are for. I take this precaution from the dual point of view of wishing to safeguard ourselves against any undesirable person being included, and also in order to make provision for the friends and relatives of the persons concerned to look after them in Palestine.

During the last few days we have received information from the following list of Russian Zionists who have been the object of political prosecutions, and either have been or will be exiled to Siberia, from which fate we can save them by the grant of visas for Palestine: Isak Rosansky, Rachil Shor, Rivka Shapiro, Fania Pipik, Aaron Brudni, Pearl Bilik. These persons have all friends in Palestine who will look after them.

Might I ask that the Norwegian Minister's enquiry receive a telegraphic reply in the following sense:

First, he should be asked to secure and transmit to us a list of the names of the 81 persons concerned, with any information available concerning them, while either he or the Political Red Cross should request the Soviet Authorities to defer action against these persons while the question of their being given visas to Palestine is under investigation.

Second: the names given above should be communicated in the same telegram to the Minister with a request that if they are among the persons applied for (as I think is almost certainly the case) visas should be issued for them.

Yours sincerely

F.H. Kisch

■ ■ ■ ■

14   H.O. Plumer,[59] High Commisioner for Palestine to L. Amery,[60] Secretary of State for the Colonies

ISA 11/1174, IMM/7 I
5 October 1927
Secret

Right Honourable L.C.M.G. Amery
His Majesty's Principal Secretary of State for the Colonies

Sir,

I have the honour to refer to my secret despatch of the 17th of August with regard to the grant of visas for Palestine to Soviet citizens under arrest on political charges in Russia.

2   Hitherto applications for visas for Palestine on behalf of persons of this class have been made by the Palestine Zionist Executive; and Madame Pechkoff [Peshkova], President of the Society for the Help of Political Prisoners at Moscow, had assisted the Palestine Zionist Executive in the matter of preparing applications and arranging the departure from Russia.

Applications have been granted on guarantees by the Palestine Zionist Executive to provide for the maintenance of individuals and its assurance as to their political good faith; and Immigration Certificates and authorisations for the issue of visas have been sent directly by the Chief Immigration Officer to the British Consulate designated. In fact, the procedure proposed in Lieutenant Colonel Symes' telegram No.106 of the 8th of June, which you approved, has been followed in all cases.[61]

3   I am now, however, of the opinion that it is advisable to amend this procedure and to treat applications in respect of Russian political refugees in the same way as applications from political suspects and to refer them to London in accordance with the procedure prescribed in paragraph 2 of your secret despatch of 21st of July.[62]

In taking this view I do not desire to impugn the good faith of the Palestine Zionist Executive in giving guarantees for the financial support and political acceptability of the persons on whose behalf it applies for visas; nor have I any reason to doubt that many of these persons are persecuted for holding or expressing Zionist views and forced to choose between imprisonment and flight to Palestine. On the other hand, means of communication with Russia are so precarious, and the sources of information upon which the Palestine Zionist Executive have to rely are so uncertain that I feel there is real danger that the Soviet Government will exploit this emigration of political refugees to Palestine as a method of introducing Bolshevik agents into this and contiguous territories.

That danger will be minimized by the new procedure which I have decided to adopt, since there will be available in London, for the scrutiny of the political character of each applicant, the records and references of the Foreign Office. I realize that this procedure will interpose additional delay in the issue of Immigration Certificates; but it is unquestionably the safest method of control in the interests of Palestine.[63]

4   I shall arrange to send with each application all evidence and guarantees procurable locally, and also photographs for the purpose of identification and as a safeguard against personation. Immigration Certificates will not, however be sent, if you agree to the suggestion in the following paragraph.

5   As you are aware, it is difficult for the applicants to make their way, en route to Palestine, to a British consulate to claim their Immigration Certificates and to obtain visas. Especially as regards Constantinople is this the case, as is evident from the *promemoria* of the Norwegian Legation at Moscow enclosed

with your Secret despatch of the 27th July and from the accompanying copy of a despatch dated the 24th of August which I have received from Sir George Clerk.

This difficulty might be overcome by the following way:

In each case in which the issue of an Immigration Certificate was approved, I should be informed and the certificate would be prepared by the Chief Immigration Officer here to be sent in due course, with a photograph affirmed, to the Immigration Officer at the port of arrival.

The applicant would be notified at the same time through the Palestine Zionist Executive that the certificate was ready. On arrival, after identification by means of the photograph, he would be admitted on a visa for Palestine granted by the Immigration Officer, on payment of the reciprocal fee applicable to Russians.

In the circumstances, I would not propose to levy any surtax. Each immigrant would still, however, require to be in possession of a valid Soviet or other national passport establishing nationality and identity, on which the port visa would be endorsed.

The Palestine Zionist Executive is in a position to notify the Government in advance of the port of arrival in each case, and to facilitate matters for all concerned; arrivals will be concentrated so far as possible in Jaffa.

6 I transmit herewith, printed on blue paper, a small portion of the required supply of forms and of the Notice for use by applicants in Soviet Russia, of whatever class, for visas for Palestine.

The remainder will be sent by the most expeditious route.

The form has been printed in English and Russian, and the Notice in Russian only.

I have the honour to be,
Sir,
Your obedient servant,

(Sgd.) Plumer, F.M.
High Commissioner for Palestine

■ ■ ■ ■

15 Correspondence concerning the substitution of Y.S. Arav and A.L. Molodetskaia-Arav (September 1926–October 1927)

GARF f. 8409, op. 1, d. 1698, l. 60
24 September 1926

To Citizen Yakov Arav[64] a.k.a. Foma L'vovich
Vinnitsa, DOPR[65]

In answer to your inquiry and in accordance with the information received from the OGPU, I can inform you that by order of the Special Commission on 18/9/1926 you are permitted departure to Palestine.

**

GARF f. 8409, op. 1, d. 1698, l. 762
12 October 1926

From: Pompolit
To: Lota Sandomirskaia[66]
Kharkov

As a result of the appeal by the polit[ical] prisoners Yakov Arav, a.k.a. Foma L'vovich, I report that his exile to Central Asia has been substituted with permission to depart for Palestine. He is travelling by convoy from Vinnitsa to Tashkent. I request that you inform him, if it is possible, when he arrives in Kharkov. We sent his notification to Vinnitsa, but it did not reach him there. There is a visa for his entry into Palestine, and we can send a notice to this effect.

**

GARF f. 8409, op. 1, d. 176, ll. 252–3
[16 February 1927][67]

Most Esteemed Ekaterina Pavlovna!

With this letter we report to you the drastic changes that have taken place in the lives of the group of exiled Zionists and Zionist Socialists in the city of Poltoratsk[68] over the last few days.

As you know, all of us in the city of Poltoratsk received substitution and our passports were to have been prepared here. The original deadline for this, 1 January, was extended to 1 February, after which they promised to send out into the localities all those who had not had time to prepare their passports by this deadline. From the beginning of January we launched a campaign for the application of the discounted rates for the passport in accordance with the social and economic status of each individual based on uniform standards, which would have made it possible for the majority to leave for Palestine in the near future. We made very slow progress resolving the issue due to unauthorized obstruction on the part of the GPU. In the end of January, after endless negotiations with all the higher state organs (the Prosecutor, Sovnarkom), we attained a resolution of the problem from the central organs which was definitively in our favour. All that remained was to coordinate this with the GPU and present certification of our social status. However, it turned out that a final resolution of this question depended on the empowered representative[69] of the OGPU in Tashkent, where our individual petitions were to have been sent. The local organs and the GPU declared their positive position on the question and promised their support. In the event that this continued even after 1 February, they officially promised not to disturb us with the prospect of sending us out to the localities. We began to prepare energetically for receipt of the passports. On 1 February at registration[70] they invited us all to appear on the following day for some sort of secret announcement. On 2 February they announced to us the new places of exile in the localities and asked us to come in the evening with our things. The question of the discounted passports was never resolved. It turned out that they had simply deceived us, although, to be sure, five comrades were allowed to stay in Poltoratsk supposedly because they would be permitted the discount passports due to their social status. There was nothing to do but accept the situation and pack our bags. The new places of exile were quite wretched, although there are even worse places here in Turkmeniia. On the 3rd [of February], the first group was sent out to Krasnovodsk: Yehuda Barer, Nison Gushanskii and Solomon Gartsman; On the 4th [of February] all the remaining groups left: to Kerki – Abram Itskovich, Sarra Shenderovich, Sema Arshavskii, Peisakh Kats and Iosif Ioshpe (the latter two are new, having just arrived); to Chardzhui – Khisia Teverovskaia; to Tedzhen – Abram Pepliker-Stern, Boris Levin, Sarra Kats, Motl Rabinovich. Remaining here are Zelik Kunin, Mania Golovei, Toiba Finifberg, Gersh Shnaiderman, Leia Gipsh for preparation of discounted passports. The rest (Yakov Arav, Boris Basin, Peisakh Krasnogorskii, Sholom Beilin and Isaak Bender) decided to prepare passports for 200 rubles and were therefore allowed to remain, on the condition that they be prepared before 5 February. Therefore this last group began urgent work to get them ready. The passports will be received in the next few days. The deadline for having the visa issued is also quite limited. Overall they have now taken quite a hard line in relation to us. We can not even manage to get out of them one extra day. We have made contact and requested money by telegraph. Now we must send out the passport in order to be issued a visa. [...] I urgently request you to clarify immediate whether or not there is a visa or certificate in our names (that is those who have begun the process of preparing the passports). Will there be any delays and what is the procedure for issuing the visa. If it is required to fill out the form beforehand and there is no way around this, as was the case with the previous passports,[71] then please do everything that is necessary. Telegraph please to let us know whether we should send the passports to you immediately or when we should send them.

Write to this address: Krylov, 3, Yakov Arav, or to be held at the post office. We thank you in advance for everything.

With respect,

Yakov Arav

P.S. If money should arrive addressed to Leia Gipsh, we request that you immediately send it here to this same address.

**

TEL: GARF f. 8409, op. 1, d. 168, l. 436
17 February 1927

Yakov Arav
Poltoratsk

Visas have been received for you and Kunin.[72] Send the passports.

Peshkova

**

GARF f. 8409, op. 1, d. 161, l. 301
5 March 1927

To the British Mission [Moscow],

I am enclosing herein the foreign travel passport for departure to Palestine of Yakov Arav (number 174045/710039) and Yakov Moiseevich Fomin (number 124725/380021) and request that visas be stamped into the enclosed passports. According to the notice received from the OGPU, the individuals in question have been permitted to leave for Palestine by a ruling of the Special Commission.[73]

[stamp of Pompolit]

**

GARF f. 8409, op. 1, d. 176, l. 261
11 April 1927

To Citizen Evelev
Odessa

In response to your inquiry I can report that a visa has been stamped into the passport of your relative Yakov Arav and it was sent to him in Poltoratsk on 28 March 1927. A telegram has been received indicating the receipt of the passport.

[Pompolit]

**

GARF f. 8409, op. 1, d. 176, l. 259–259ob.
27 April 1927

Most Esteemed Ekaterina Pavlovna!

I am hereby informing you that the last group of exiled Zionist socialists and Zionists from Poltoratsk, numbering nine people, arrived in Odessa on 23 April 1927. The steamer to Jaffa is leaving on 5 May. Allow me once again to thank you warmly in the name of the entire group for the assistance you have provided through the entire period of our stay in prison, convoy, exile and finally in the process of our departure to Palestine. Your help has been invaluable. Accept our greetings and best wishes.

And now, I have one more personal request for you. On my arrival in Odessa, I was married here to my Comrade A. L. Molodetskaia. She is not an exile, but it is essential that she leave the country with me or on the following steamer. She does not have her own visa. As far as I know, the husband's visa can be applied to the wife's passport. My visa is valid for entry until 31 June. But I am still listed at the consulate as single. It is unclear, therefore, what exactly needs to be done in order that my visa be stamped into her passport which she will send you in the near future. Will we need a notice from ZAGS[74] or some other kind of certification, or perhaps my last name is enough if it will appear in her passport? I urgently request that you regard this problem in all its gravity, as [you do] with [the problems of] political exiles in general, taking into account that this problem is quite urgent and can not be put off. If there are certain complications in entering my visa into her passport, although it seems there should not be, then I ask you to do whatever it takes to resolve them. Her passport will not be prepared until detailed information is received from you of the possibility and procedure for processing it with a visa. Therefore, I ask you to respond to all my inquiries immediately at the address: Odessa, Main Post Office, to be held for A. L. Molodetskaia.

In full confidence that you will provide sufficient assistance.

Y. Arav

**

GARF f. 8409, op. 1, d. 176, ll. 256–256ob.
3 May 1927

Much Esteemed Ekaterina Pavlovna!

I have not yet received an answer to my message from 26 April 1927[75] regarding the entry of my visa into my wife's passport. However, your answer to Comrade I. Bender[76] regarding an analogous case can serve as a guide for me as well. Therefore I am sending you with the present letter my certificate (number 1922) and marriage license issued by the Odessa ZAGS on 30 April 1927 (number 1160), as you have indicated, to be presented to the Mission for the necessary notations. The ship on which I must depart leaves on 12 May 1927, i.e. in nine days. Of course, I will need the certificate for departure. The slightest delay in sending it back to me threatens me with misfortunes: (1) my passport will expire and be lost; (2) I will be sent back to prison. Therefore, I urgently request that you make the necessary notations at the mission immediately upon receipt of the certificate and then send it back to me immediately. If for some reason this notation cannot be made immediately then you should of course not delay the certificate, since I will need it here several days before my departure.

My wife, Molodetskaia, has already ordered her passport and will be able to send it for visa processing in the next few days. Please do everything you can so that there are no delays in the processing. Send the certificate to me at the address: Odessa, Konnaia 13. So, I hope there will not be any delays or problems with the certificate.

I thank you in advance.

Respectfully,

Yakov Arav

P.S. [In handwriting]

Esteemed Citizen Peshkova!

My husband took upon himself the risk of sending you his certificate when only a few days remain. His passport was to expire on 15 May, and it took extraordinary effort before they extended it for seven more days. Now he has sent off the document without which he will undoubtedly not be able to depart, without thinking that it could be late by a single day and then all the ordeals, and the suffering and his last money, all this would go to waste and he would face prison. Therefore, I ask you to do everything possible for me, but most of all to send back the certificate immediately. We have counted not only the days, but the minutes and if you send it on Saturday, then it will be on time, even if it is Saturday evening, but not later. I ask you to take all of this into account and to send it right away. If my husband were in good health then I would not have taken on such a risk, but I must go, if not with him then to him, because he is ill. I urge you to send everything back right away.

Molodetskaia.

**

TEL: GARF f. 8409, op. 1, d. 176, l. 251
11 May 1927

Odessa
To Arav

The certificate has been processed. Today it was sent out by express [mail]. There is no air [mail]. Try to get an extension.

Peshkova

**

TEL: GARF f. 8409, op. 1, d. 176, l. 248
11 May 1927

Moscow, Pompolit

It is completely impossible for me to stay. I am leaving. Send the certificate by air [mail] to Palestine. Let the consul send a telegram to the port of Jaffa so that they let me off the ship. Let Pines send a telegram to the Histadrut to come and get me.

Arav

**

GARF f. 8409, op. 1, d. 176, l. 246
13 May 1927

To the British Mission [Moscow]

Yakov Arav, who has received a passport for foreign travel stamped with a visa and a certificate for departure to Palestine, sent his certificate so that an entry could be added regarding his marriage (in light of the fact that his wife is also departing for Palestine in the near future). The certificate with your notation to the effect that Yakov Arav is married to Anna Molodetskaia was sent to him by express mail, but at the time when the ship was to depart on 12 May 1927, the certificate had not been received. In light of the fact that the passport was to expire, Arav could not delay and departed. We request that you not desist from sending the necessary telegram to Palestine (at his expense) so that he will be allowed to disembark.

[Pompolit]

**

TEL: GARF f. 8409, op. 1, d. 176, ll. 239–40
16 May 1927

Odessa-Jaffa, Steamer *Tobol'sk*
To Passenger Arav

According to the consul [you] will be allowed into Palestine. In the event of being detained, telegraph.

Peshkova

**

GARF f. 8409, op. 1, d. 176, ll. 239–40
20 May 1927

Most esteemed Ekaterina Pavlovna!

Unfortunately, I again have to turn to you and ask for your assistance and advice in my affairs. The problem is that when my husband left it seemed that my passport was already almost in hand. Today I received a letter from Uman' where they write to me that my situation is complicated in the following respect: My husband and I had very little money. All that I had left was money for the ticket and 95 rubles for the passport. I received a certificate from my trade union releasing me from paying fees to all the institutions to which I was obliged to turn. I did not go there myself since I did not have the extra 20 rubles for the trip. I had to make do with a power of attorney made out to an acquaintance, the results of which were as such: being almost indifferent to my situation, he paid everywhere. This business cost me 20 rubles and after all of this they answered him at the administrative department that I would have to receive a passport in Moscow since, beginning in 1922, I lived in Moscow until 1926, and from the time that my husband was arrested I began to drift from one relative to another. Now I have lived in Odessa for three months already. In short, since 1926, I have not had a permanent residence, therefore I must receive a passport in Moscow. Very little time is left for me – less than three weeks, in the course of which, I will not under any circumstances be able to arrange a passport since in Moscow I would have to wait six weeks at the administrative department alone. My visa expires on 1 July which means I can get on the last steamer on 10 June. What am I to do? I turn to you as the only person who remains for me here. After all, if I do not go now, then what is left for me. My husband is still not a British subject, he can not send me a visa. I have absolutely nowhere to live here. What am I to do now? I ask you to take into account my hopeless situation and, if it is possible, to somehow speed up this matter through your intercession. Inform me. Either I go [to Moscow] if it is necessary, or I will send to you the necessary papers: the birth certificate, marriage license and proof of membership in the trade union to receive the discount passport and, if it is necessary, the papers issued by the Uman' institutions. I repeat, if it is necessary, I will come to Moscow, but before that I ask you to respond to me about what to bring, which papers. I await your answer upon which the future depends.

Respectfully yours,

Molodetskaia-Arav

P.S. I sent the certificate received from my husband right away by air mail. Today I got a telegram from my husband from Constantinople. They are sailing onward today. Write back to: Odessa, Main Post Office, Hold for A.L. Molodetskaia.

**

GARF f. 8409, op. 1, d. 176, l. 238
21 May 1927
Odessa, Main Post Office
Hold for A. Molodetskaia-Arav

In response to your appeal, I can report that we will try to obtain a passport for you here. For this you will need to send all your papers here and the money for the passport and also fill out and send back the attached form of the AOMS[77] along with two photographs. We will inform you of the outcome by telegraph. We will try to extend your visa. Send your papers by express mail.

[Pompolit]

**

GARF f. 8409, op. 1, d. 176, l. 237
3 June 1927

To British Mission [Moscow],

Please take measures to see to it that Anna Molodetskaia, the wife of Yakov Arav who has departed for Palestine, and whom you entered onto certificate number 1922 issued to Y. Arav, is permitted to disembark from the ship in Palestine at a date somewhat later than the date which you indicated in the above-mentioned certificate (i.e. after 30 June), in light of the fact that there has been a delay in the receipt of her foreign travel passport.

If you find it necessary, I request that you send a telegram to this effect to Palestine at her expense.

[Pompolit]

**

GARF f. 8409, op. 1, d. 176, ll. 231–2
24 September 1927

Most Esteemed Ekaterina Pavlovna![78]

After very long and zealous efforts with the Emigration[79] Department of the Palestine Government to get them to send out a visa to Constantinople for my wife Anna Molodetskaia, I was finally able today to receive official notification that an Immigration Certificate numbered 3529 has been sent out in her name to the British mission in Constantinople.[80]

Attached, I am sending you this notification in English and I ask that you take care of the question of the Turkish visa for my wife so that she can leave immediately and stay in Constantinople while the Palestinian visa is stamped into her passport. If this question can be taken care of in some other way, then by all means take care of it.

Please also inform my wife immediately about this notification, so that she can leave right away, since the passport she received will expire if it has not already. Just in case, I will give you her address again: Moscow, 3 Lechebnaia St., Apt. 4, Anna Molodetskaia. Now there are no more obstacles to her departure. I ask you once again to provide her active assistance so that she can depart immediately, if such help is needed, since she has had to endure too much. And if she does not hurry now, then her passport will be lost and then the entire process will have to begin again.

I thank you in advance for the assistance.

With socialist greetings,

Yakov Arav
Palestine (Nahalal)[81]

P.S. To avoid any kind of misunderstanding, please pass on the notification to my wife, just in case she needs it as an official document when she goes to have the visa entered.

Arav

**

GARF f. 8409, op. 1, d. 176, l. 230
21 October, 1927

Odessa, Main Post Office
Hold for A. Molodetskaia Arav

In response to your inquiry I can inform you that in accordance with the notification received from Palestine, a visa (number 3529) has been sent to Constantinople in your name for entry into Palestine.

[Pompolit]

**

GARF f. 8409, op. 1, d. 176, ll. 228–9ob.
27 October 1927

Esteemed All,

I want to stand by my promise and keep you informed. Approximately 10 of us are going. I saw Ioshpa today. He is not going, he has some problem with a visa and based on my experience, I think that this is not a problem that will be resolved anytime soon. Marinov is also not going. That is as much as I know now. I will report en route whatever I find out about the rest. Did they receive the grammar book? That is all for now. Stay well.

Arav [Molodetskaia]

P.S. Ioshpa telegraphed to Jerusalem. There is sill no answer. I saw Marinov. He is planning to leave.

■■■■

16   Ch. Halperin, Immigration Centre to Pompolit

GARF f. 8409, op. 1, d. 232, ll. 34, 234-ob.
Tel Aviv, 14 June 1928

To the Committee to Aid Political Prisoners
Moscow

We have received 205 English pounds designated exclusively for travel expenses to Palestine for the following individuals:

1   Boris Rubinshtein – Ust'-Sysol'sk – 15 pounds
2   Mikhail Shekhter – Khorezm, Khiva – 15 pounds
3   Moisei Shtekelis – Simferopol' ? – 10 pounds
4   Liza Kaminer – Tobol'sk – 10 pounds
5   Leizer Trakhtenberg – Taldy-Kurgan – 10 pounds
6   Moisei Finkel'berg – Obdorsk – 20 pounds
7   Yankel Markman – Chimkent, 52 Sadovaia St. – 15 pounds
8   Il'ia Gushanskii – Izhma – 10 pounds
9   Mania Pechenik – Kokchetav, Akmolinsk Guberniia, 5 Pushkinskaia St. – 20 pounds
10 Sonia Bakaleishik – ? – 15 pounds
11 Yakov Krutitskii – Karakala, Kzyl-Arvet, Turkmen SSR – 20 pounds.
12 Khaia Vilenskaia – Karakala, Kzyl-Arvet, Turkmen SSR – 15 pounds.
13 Zina Rodionovskaia – Ashkhabad, 2 St. Pervomaiskaia – 20 pounds.

Total: 205 English pounds.

Today we sent out money through two couriers: Mrs. Gnesin (30 English pounds), and Mrs. Cherniak (60 English pounds). In the next few days we will send out 105 English pounds. Upon receipt of the entire sum, please inform the exiles named above that you have for each of them the indicated sum, which you will give out to each under the condition that they have available the base sum necessary for travel, i.e. the sum you give out should be used as a last resort to cover any shortfalls.

In the event that someone from among the above-mentioned exiles is in need of a sum that is less that what we are sending we request that you hold back the remainder and at the conclusion of the entire operation inform us, how much was transferred, to whom, when they are leaving and the amount of the remaining sum.

We need as much detail as possible for our record keeping. We request that you collect signatures indicating receipt of the funds from each of the recipients and send to us all 13 receipts in the format that is customary for your organization.

There is no hope of obtaining additional sums for these individuals, and therefore we request that you inform us who can not make use of the sum offered (if it is not needed or not sufficient), so that we might indicate in a timely fashion other names to whom the money should be transferred in such cases. Of the 205 pounds we are only sending 195 directly to you since we gave out 10 pounds today to the sister of Kaminer to be sent to Tobol'sk.

The addresses of the exiles that we have indicated are incomplete and imprecise: please check them and make corrections.

[signed]
Halperin

■ ■ ■ ■

17   Pompolit to Immigration Centre

GARF f. 8409, op. 1, d. 232
12 December 1928

To the Immigration Centre
Palestine, Tel Aviv

We confirm the receipt of your letter of 22/11/1928 with the attached list of visas (35 [in number]) and the five visa certificates.

In your letter of 2/8 of the current year, you indicated that in the event that Natan Ekhevich, living in Tiumen', is permitted to leave for Palestine, the money needed for his departure would be sent to our address for him. According to a notice we have received, his departure has been permitted. We have already informed him of this.

Please let us know whether the visas which were due to expire on 31/12 of this year were extended, and if so by how much; please inform by telegraph about the possibility of entry with the aforementioned visas.

Please clarify and inform us about the possibility of entry of wives separate from husbands when they have a family visa. Can they come either before or after the arrival of their husbands? In particular, what is the status of the question of a visa for Maria F. Gushanskaia whose husband, Ilya Khononovich Gushanskii, had a family visa but has already left for Palestine. The wife, because of the lack of funds, was delayed. Can she enter on the previous visa number 4168, or will you now need to petition for a separate visa for her?[82]

If this is impossible now, then please ascertain whether it is possible to receive anonymous visas in order to accelerate and ease the departure for those who received the substitution of the sentence with departure to Palestine.

Please petition for visas for the following individuals:

1 Vol'f Samuilovich Chechik, born 1876, living in Troitsk, his wife, Gisia Saulovna, born 1876; mother – Musia Khatskelevna Paperno, born 1853; his daughter Liia Vol'fovna, born 1916; son, Samuil, born 1918 and son Saul Vol'fovich, born 1919.
2 Moisei Davidovich Gornik, born 1895, single, living in Kassan.
3 Zinaida Davydovna Fridman, born 1910, unmarried, living in Aktiubinsk.

■■■■

18 Central Office of the Zionist Organization (London) to members of the Executive Committee and to the Zionist federations and fractions

Lavon Archive IV-104-53, f. 35
4 March 1929

Dear Comrades,

Re: The Condition of the Zionists in Russia

The condition of our persecuted, suffering Zionist comrades in Soviet Russia has long provoked great concern among Zionist circles throughout the Jewish world. On countless occasions we have sought to come to the aid of our comrades in Russia who, despite the dangers, have worked ceaselessly and energetically on behalf of Zionism. They have sustained the Zionist tradition with the utmost devotion. To our great consternation, we have to confess that our efforts on their behalf, which included an attempt at the last Congress in Basel to establish an aid committee for those being persecuted in Russia, have brought about no tangible results. The Executive makes every possible effort within its budget to extend financial support for Zionist work in Russia. But these activities have failed to give the assistance that is required by the Zionist movement in Russia and by our persecuted comrades there.

This indifferent attitude on the part of Zionists who are fortunate enough to live in better conditions, must come to an end. We must, without delay, fulfil our moral obligation towards our comrades in Russia and prove to them through our actions the solidarity of the Zionist movement.

The Executive appeals to the Federations and Factions with a request that they participate in the creation of a *special fund* amounting to 2,000 pounds sterling. Part of this fund will be devoted to assisting the immigration of Russian Zionists to Palestine. Part will be used to improve the living conditions of political prisoners. And part will go to the work of the [Zionist] organizations.

We are sending you, *for your eyes only*, a memorandum composed by a Zionist with knowledge about current conditions in Russia that describes the situation of our imprisoned comrades. The conclusions presented at the end of the memorandum are the private opinions of the author and need not be taken into consideration.

114 *Documents*

What is most important is the faithful description of the present situation in Russia. Who could read this report and not be moved and troubled by the hellish suffering our comrades in Russia are forced to endure?

All our Federations are requested to pay close attention to our letter and the attached memorandum. Our desire is that you convene your central committee to meet in a special session in order to discuss the situation in Russia in a serious and sympathetic manner. We believe that it will be easy to collect the required sum of 2,000 pounds sterling, on the condition that two activists from each country assume the task. As an exception, a number of affluent Zionists should be approached with the aim of bringing this project to an early conclusion, and that by means of several extraordinary donations. It will be easy to obtain contributions for this cause – which is at once national and humanitarian – even from non-Zionist circles, and from general Jewish organizations and communities. The attached list contains the share that has been assigned to your country for raising the sum of 2,000 pounds sterling. The allocations presented herein are minimum amounts. Wherever it is possible to surpass these figures, the opportunity should be used to raise more than the sum that appears here.

We turn to you and call upon you to extend assistance to our Zionist comrades in Russia. The Zionist Executive has a moral obligation, and it is up to you to make it possible for the Executive to fulfil this obligation.

We await acknowledgement of the receipt of our letter in the form of a promise that you will do all that you are capable of in order to fulfil this request.

With great respect and Zionist salutations,

The Executive of the Zionist Organization
[SIGNED] Felix Rosenblit, Member of the Executive

Supplement[83]

Contributions towards supporting Zionism in Russia
Contribution shares by individual country

	£
Eretz Yisrael [Palestine]	50
America	500
Argentina	50
Australia	25
New Zealand	25
Austria	25
Belgium	50
Bessarabia	15
Bucovina	15
Brazil	10
Bulgaria	25
Canada	250
Chile	10

Continued

	£
Czechoslovakia	50
England	125
Estonia	25
Finland	25
France	100
East Galicia	25
West Galicia	25
Germany	200
Holland	50
Hungary	20
Italy	20
Yugoslavia	15
Latvia	50
Lithuania	30
Poland	70
Romania	85
South Africa	250
Switzerland	20
Transylvania	10

On the persecution of Zionists in Soviet Russia

The persecution of Zionists of all stripes and party affiliations in Russia, which began in an organized fashion in 1922 on the part of the Soviet government, has been going on for seven years. During this time the cream of the membership of all the various Zionist movements has been arrested and incarcerated. Hundreds have been deported to distant regions of northern Russia, Siberia, Turkestan, etc., or have been sentenced to lengthy imprisonment in *politizoliatory* (special penitentiaries designed for political criminals). At present, about 1,500 members of various Zionist parties are to be found in Russian jails, in *politizoliatory*, or in exile. Hundreds of comrades have received permits to immigrate to Palestine after years of suffering and affliction in prison and in exile. The weaker – those who failed to sustain and control their spirits and who lacked the courage to continue the struggle – have abandoned the battle. Only those Zionists who have remained loyal to the ideal of national resurrection, modest numbers of comrades who have martyred themselves for the Zionist cause, continue their Zionist activity to this day, in spite of being surrounded by stalkers and oppressors, who are conspiring daily to undermine the movement's existence. The GPU continues to find among these surviving refugees sufficient reserves of candidates for imprisonment and exile.

The past year, 1927–28, is no exception in this respect. In 1927, and in particular during the early part of 1928, a considerable resurgence of Zionist sentiment among Russian Jewry was in evidence. There were several causes for this: the difficult economic situation in which the masses of Russian Jews find themselves – the likes of which has not been seen for a long time – resulting from changes in

Soviet economic policy which have been to the detriment and subjugation of free trade, and owing to the partial resuscitation of War Communism; the growing level of antisemitism, which increasingly penetrates communist circles despite the efforts by upper party institutions to arrest this phenomenon; the halt to settlement in the Crimean Peninsula and the great disappointment that followed after so much hope was invested in it; and, most especially, the reports concerning the improvement of the economy in Eretz Yisrael.

All these factors have naturally given birth to an oppositionist temperament and have reinforced the objections of the masses of Jews to the Evsektsiia and the oppressive measures undertaken by the Soviet regime. Together with these developments, there has been a significant awakening and, as far as is known, renewal within the memberships of various Zionist organizations, particularly that of Hechalutz. At the same time, the guardian of communism in Russia – the Evsektsiia – neither rests nor sleeps. In order to suppress this renewal before it can develop, the Soviet government has begun to strengthen and revive its persecution of Zionists, whom it views as its principal competitor and opponent amongst the Jewish public.

It is worth noting that if, up until last year, the principal effort was directed against the political Zionist organizations (Hitachdut, the Zionist Socialist Party, the Zionist Youth Organization – EVOSM, and Ha-Shomer ha-Tza'ir), the GPU has stepped up its persecution of the Zionist pioneering movement in the past year,[84] with the aim of bringing about its total liquidation.

This was most evident when Legal Hechalutz was dismantled at the beginning of 1928 (as is known, since 1923 a second Hechalutz organization has existed in Russia which is known by the name, Illegal Hechalutz). This action revealed that the real intention of the Soviets was to root out any possibility of Zionist activity in Russia. Whereas the GPU persecutes the political Zionist organizations by accusing them of purported anti-Soviet activity, there was no basis for levelling any such accusation against the Hechalutz movement, and especially not against Legal Hechalutz, most of whose members belong to no party, engage in no political work, and are principally devoted to training for a life of labour in Palestine.

But not even the obvious loyalty of Legal Hechalutz to the Soviet regime was of any help when the dire hour arrived. It could not rescue the movement from destruction. In addition to publication of the order directing the liquidation of Legal Hechalutz – both the central organization and its branches – the largest farm built by Hechalutz in Russia, Tel Chai, was also terminated. A special committee headed by Merezhin[85] himself, director of the Evsektsiia, visited the farm to see to its obliteration. The large majority of the farm's members were dispersed and its committee members were arrested. After the break-up of Tel Chai the liquidation of other training institutions followed.

The condition of Illegal Hechalutz is also deteriorating from day to day. The large majority of its farms in Podolia and other locations have been destroyed and dismantled in recent years by the GPU or by the Evsektsiia. Hundreds of comrades have been deported to northern Russia and Siberia. The great devotion of the comrades has made it possible until now to protect several of the projects of Illegal Hechalutz, in particular the large farm located in the Crimea, Mishmar, which at one point had 140 persons working on it. In spite of the fact that the farm is not officially connected to the Hechalutz organization, it has been the object of constant pressure for several years now on the part of the GPU. Over the past year harassment on this farm has become systematic and persistent. Frequent arrests are made among members of the farm, almost once a month. Members of the farm directorate are in particular targeted for arrest and deportation. A while ago eight members were arrested, among them the head of the farm. Newcomers are unable to gain admission to Mishmar unless they do so through the Komzet (the communist organization for Jewish settlement in Russia). By means of systematic persecution and constant pressure, including economic pressure, and by denying credit to the farm from governmental agencies, the GPU intends to bring an end to Mishmar, the model farm of Hechalutz in the Crimea which has served for several years to attract young men and women to training for a life of labour. Mishmar is also expected to become defunct in the near future.

The pioneer movement in Russia is, thus, now a principal target of pressure by the Soviet government and its servant, the GPU, in their war against Zionism. The Evsektsiia, the agent of the Soviet government amongst the Jews, cannot come to terms in any way with the growth of support for Eretz Yisrael among the Jewish masses and youth, and among the Jews in the new farming settlements. Its war against Zionism is a determined one.

Alongside the growing repression of the Hechalutz organizations, the usual persecution of the various Zionist organizations also continues. On 12 February of this year, 18 Zionists were arrested in Moscow, among them several of the leaders of the Hitachdut Party and Illegal Hechalutz. Two of these comrades, who were held in the worst possible conditions, were forced to declare a hunger strike that continued for 11 days. Some of those arrested were sent into exile, while the leaders were imprisoned in *politizoliatory*. In early June, arrests of Zionists were carried out in a variety of locations in Russia. More than a hundred were incarcerated. As in the past, but especially more so recently, the number of those arrested who are under eighteen years of age is increasing.

The arrests took place not only in Jewish settlements, but in cities located in central Russia and in Siberia as well, areas where the Jewish population is not numerous.

The condition of prisoners and deportees

Investigation of the crimes of the Zionist prisoner usually lasts 3 to 4 months. During this time the prisoner is held in the region where he was arrested. His condition is usually very difficult. In most cases the Zionists are incarcerated in the same cells that also hold murderers and other criminals. The irresponsibility of the prison administration knows no bounds. Lodging complaints before the authorities is pointless. Zionists are beaten just as criminal inmates are. This occasionally reaches the point of horrible blows, as happened in Kremenchug in December 1927, when two Zionists were cruelly beaten for no reason in room 12 of the prison there.

After the harsh tortures suffered by the prisoner over the first three or four months – until completion of his case – the GPU headquarters in Moscow pronounces the punishment that is to be meted out to him. This is done administratively, without his crimes having been explained in a trial where he was present. (As is known, the Soviet government has long ceased holding trials for Zionist criminals since they have no actual proof that Zionist activity is subversive to the government.)[86] The punishment is usually deportation for a period of three years to northern Russia, to Siberia, to Turkestan, etc. Zionists who are accused of having filled responsible roles in the movement are held in *politizoliatory* for a similar period [of three years].

In the regions of northern Russia, the Urals, Siberia, and Turkestan, in places far removed from settlement and located hundreds of kilometres from the railroads, in hidden corners that have no contact with the outside world for most months out of the year, amongst semi-savage peoples and tribes – here hundreds of Zionists are dispersed, most of them young, living a life of want, poverty, loneliness, and idleness. Only in few instances does a professional amongst the deportees find some temporary work.

How can one begin to recount the tortures of body and soul, the disease and the physical ruin, the failing emotional energy and the intellectual deterioration that overtake the exile and the prisoner during the years of deportation or of incarceration in the *politizoliatory*!

We have a collection of letters from the sites of deportation that give an account – as much as is possible under the conditions of Soviet censorship – the tragedy and adventures of the Zionist exiles. A letter written by a deportee in the Kirgiziia district on 10 March of this year contains a description of the conditions of existence under exile. We will quote several passages from it:

> I was surprised to receive one and a half pounds (which is equal to 14 rubles and 16 kopeks in our money). The money comes just at the right time. I need it for bread and potatoes: I can't desire more than this since life doesn't offer more. I'll describe to you a day in my life here, after which you'll understand everything, since every day is just like the next. My room measures 4 by 4 [metres], the rent for which is 6 rubles a month. The monthly 'assistance' offered by the GPU amounts to 6.25 ruble. I rise at ten or eleven in the morning in order to save on breakfast. If I've got potatoes in my possession for lunch then it's a happy day, for I live mostly on bread alone. I appear at the GPU in order to register each and every day. For weeks on end, I and my comrades here are looking for a day's work. However, all our efforts are useless. There is no work to be found....

A letter similar to this one was received on 6 March 1928 from the other end of Russia, from Krasnovodsk, which lies on the Caspian Sea. It too describes the tragedy of healthy persons who have been sentenced to a life of idleness that over the years ultimately brings about intellectual deterioration and physical frailty.

The condition of frail and ill comrades is many times worse. The GPU does not take into consideration any physical ailments that afflict the deportee: neither heart disease nor tuberculosis can alter the horrible sentence imposed by the GPU, which treats the comrades in an arbitrary manner offering no justice or a hearing. The condition of those in exile who fall ill is described in the following letter, which was received on 12 January from the remote village of Vikulovo, which is in the Ural district, and which we will now quote directly:

> Comrade G., who lives here with me, developed a serious throat infection. His condition was dangerous but in spite of all efforts, the GPU refused to transfer him to a hospital in the nearby district city. We endured a lot because we had to stay up with him night and day. Now, as he recuperates, we have no way of supporting him. His illness made our rather hard situation even more difficult....

Prisoners in the *politizoliatory* as well do not receive the medical assistance they require when they are taken ill. In the Tobol'sk *politizoliator*, for instance, there is no physician on hand, only a paramedic. A physician can only be requested in extraordinary instances. Those ill with contagious diseases (tuberculosis in the final stage, etc.) live among the other prisoners [...]

The comrades often declare hunger strikes as a protest against the absence of sufficient medical assistance in the prisons and in order to press for the improvement of general prison conditions. In 1927 a group of Zionists in the Sverdlovsk *politizoliator* declared a hunger strike that lasted 14 days. In March 1928 a group of comrades in the Tobol'sk prison went hungry for 9 days. Close to a majority of the Zionist prisoners and deportees have at various times attempted to bring about an improvement in the conditions of their incarceration by means of an extended hunger strike. However, the effort usually fails to bring about the desired effects.

The victims of deportation are many. Numerous comrades have died from disease, from hunger, and from the difficult living conditions in general. One of the most recent victims was Comrade F. (a member of the Zionist Youth Organization – EVOSM) who died of tuberculosis in a village called Kurtamish, which is located in the Kurgan district in the Urals.[87] One of the comrades writes the following regarding this tragedy:

> Comrade F. fell ill back in 1926. She was sick intermittently until September 1927. From September until the day she died she did not rise from her bed. After persistent and strenuous efforts on her behalf, permission was granted in December 1927 to transfer P. to the sanatorium in Sverdlovsk at her own expense, but it was too late. Last summer she thought she would immigrate to Palestine but the physician forbade travel by sea. It is possible that the climate in the Crimean Peninsula could have saved her but the authorities refused to release her, and even refused to grant her an amnesty in the final days preceding her death, notwithstanding the fact that she was suffering from tuberculosis and heart disease and had already served three years of her sentence. She died on 24 January 1928.

The Zionist deportees occasionally experience tragedies that are not the result of objective conditions but, rather, the tricks of the GPU. One such incident occurred this year in a distant town located in Khiva (Turkestan), populated by half savage Turkmen. About 20 Zionist members of various parties have been deported to this town. In early February, all of them were suddenly arrested. A letter from one of the comrades, dated 12 April from this year, provides details of what transpired:

> Early on the morning of the first of February of this year arrests were made and all the comrades deported to Khiva were rounded up. Everyone was told that they were being accused of violating penal code 71 (anti-Soviet propaganda). After tormenting us, each one of us was locked up in a separate room. On 17 February, during the regular exercise period in the prison yard, one of the guards shot and killed Comrade N.B., doing so without any provocation and without a warning of any kind. The comrade died on the spot. Only on the next day, after they had dissected the corpse, were we allowed access to the morgue. We were witness to a horrible scene. N. lay on a wooden bed covered in blood, his head sliced open. His face had a faint smile on it since he had been killed suddenly. ... We buried him alongside another comrade who died of malaria here on 27 September 1927. ... I will make do with these few lines. There are many other things that are difficult to write about in a letter. There is only one more thing I must note: the lawlessness and the lack of any punishments for these tormenters create conditions that are impossible to live under. Who knows what the future holds for us.[88]

This tragic incident in which Comrade N.B., a member of Ha-Shomer ha-Tza'ir, was murdered sent shock waves throughout all the communities of deportees, instilling in them fear of what the reigning lawlessness would mean for them.

One can keep adding to the list of tortures experienced by Zionists from all the movements: the beatings suffered by Zionists in the Kharkov prison in August 1926 and the sentence of five years in the *politizoliatory* meted out to those comrades after the beating; the dispersion of comrades deported together to Narym (Siberia) and the transfer of dozens of comrades to *izoliator*, the dispersion of those who were in Irbit, and, most recently, those who had been deported to Kazakhstan and their transfer to even more distant locations; the sudden transfer of comrades from Eniseisk (north-eastern Siberia) at the end of June of this year to the most gruelling, remote locations in terms of climate. The suicide of Comrade A.S. (Illegal Hechalutz) in Chimkent (Kazakhstan);[89] the instance of the suicide of Comrade Z. in Ekaterinoslav (EVOSM) after the GPU forced a declaration out of him denouncing Zionism and Zionist activity; the arrests of Zionist youths 16 and 17 years of age and their deportation to the Urals, Siberia, etc.

The suffering and the tortures experienced by the comrades do not end even after they have served their sentences, sitting for three years in their place of exile. These comrades are then denied permission to reside in Ukraine, White Russia, Moscow, Leningrad, Kiev, Odessa, Kharkov and Rostov on Don, that is to say, in those places with high concentrations of Jews. As a consequence, the comrades are forced to find themselves a place to live in central Russia such as Voronezh, Kursk, Saratov, Kazan, etc. They are not allowed to then leave these places without authorization from the GPU. For all practical purposes, this constitutes a continuation of their deportation. Their material situation is most difficult as well since they cannot secure employment as deportees. In addition, they are liable to fall prey at any moment to new persecutions and renewed arrest and deportation.

Conclusions to be drawn from the above report

1 Responsibility for the persecution of Zionism and of the Hebrew language in Russia not only falls on the Evsektsiia, but on the Soviet government in general. (It is worth noting here that in many places, those members of the GPU assigned to investigate Zionist crimes are often non-Jews.) The war against Zionism is no less determined or cruel than the war waged against the counterrevolutionary parties. The escalating persecution of the Zionists is designed to destroy the Zionist movement by means of both the physical and spiritual oppression of its followers.

2 In order to protect the remains of the Zionist vanguard in Russia, so it is not wiped out forever, and in order to defend the new Zionist vanguard that is still young and lacking in tradition, and to ensure that it not be nipped in the bud, it is necessary, in addition to offering material assistance, to do the following: improve the regime in relation to the Zionist prisoner and deportee; ensure sufficient medical attention; get the Zionists out of the most distant locations with the worst climates from which no one ever returns; increase governmental support to the minimum required for subsistence; cancel the high cost charged the Zionist deportee who receives an immigration permit to Eretz Yisrael for a passport.

A general survey of the situation cannot fully describe the extent of the persecutions, the hardships, and the terrible suffering experienced by the Russian Zionists of all stripes and parties. Most probably, a large percentage of the tortures of both body and soul endured by the deported prisoners are not even known because of the great distance, the censorship, and the oppression practiced by the GPU, which maintains a close watch over the 'Zionist criminals'. We are faced with a great human tragedy affecting hundreds of persons who are waging a selflessly brave struggle on behalf of the Zionist idea and Zionist deed. Their struggle is not accorded the proper respect within the world network of Zionist activity maintained by the Zionist Organization and its leadership. There is no certainty that the Zionist movement in Russia will have the internal strength to hold on in the future under such conditions of constant external pressure – unless the World Zionist Organization immediately acts to extend tangible assistance to the Zionists of Russia who are doing battle and are being sent into prison and exile, and, together with that assistance, pursue a policy of political intervention in order to relieve their situation and suffering.

■ ■ ■ ■

19 A. Saunders, Department of Police and Prisons, Government of Palestine to Chief Immigration Officer

ISA 11/1174, IMM/7 II
Confidential
Jerusalem, 12 April 1929

Department of Police & Prisons
Chief Immigration Officer

Subject: Grant of visa for Palestine to Soviet Citizens

Under No. 2135/29 the Chief Secretary has approved of the following measures in respect of above:

(a) when an application is received by you from the Palestine Zionist Executive you shall require the latter, who probably has means of confirming that the applicant is a Zionist refugee or is not connected with the communist, Left of the Po'alei Tziyon or kindred movement;

(b) you shall require applicants for traveller's visa or Immigration Certificates, other than those submitted by the Palestine Zionist Executive, to give references in Palestine in all cases where this is possible;

(c) when the application is not received through the Palestine Zionist Executive, or when the latter has declined responsibility, a certificate as required under (a) shall be accepted from the references in Palestine, if any, only within the discretion of the Commandant and the Chief Immigration officer;

(d) under Section 5(2) of the Immigration Ordinance you shall require an immigrant from Soviet Russia to make a deposit or give a bond to cover the expenses of his deportation, in case he should be found to be an undesirable. If the would-be immigrant obtains a bond from a surety instead of giving it himself, the Government would be equally protected. The form of bond may also be required in respect of all persons for whom the Palestine Zionist Executive applies except such as are certified by the Palestine Zionist Executive to be political refugees, since it would be impossible to deport such persons to Russia. A bond should only be required from travellers in respect of whom doubts are entertained;

(e) for the time being and until there is more control of communist propaganda, by disposal of the present heads of the movement and by providing more adequate means to control extremist tendencies, immigration from Russia shall be stopped, except in the cases of persons certified in accordance with the procedure laid down in paras. (a), (b) and (c).

[…]

8 The following should be the procedure in dealing with these applications: When an application has been submitted according to the approved form and the applicant has satisfied the preceding conditions, reference shall be made simultaneously to this office and to the Home Authorities. When a reply is

received from the Home Authorities it shall be notified to the Commandant who will then communicate his decision as to the admission or otherwise of the person concerned.

A. Saunders
Commandant

■ ■ ■ ■

20 Palestine Zionist Executive to the Chief Secretary, Government of Palestine

CZA S6/f. 5355
10 June 1929

The Chief Secretary
Government Office
Jerusalem

Sir,

I have the honour to transmit herewith two lists containing the names of 46 Zionist political refugee (33 men and 13 women) and to request you to be so good as to authorize the Chief Immigration Officer to issue Immigration Certificates in their favour after making the usual enquiry.

    2   The lists give also the names of the parties to which the refugees belong, so far as this is known to us, in accordance with the terms of your letter of the 4th inst.[90]

    3   The Palestine Zionist Executive guarantees the maintenance of the refugees whose names are given in the attached list for a period of one year from the date of their entry into Palestine as well as their political bona-fides.[91]

I have the honour to be,
Sir,
Your obedient servant

Palestine Zionist Executive

Name	Age	Previous residence	Place of prison or exile	Party
Men				
1. Askinasi Mordechai	29	Leningrad	Surgut	Hitachdut
wife Miriam	23			
2. Baril Meir	21	Olevsk	*etap* [in transit]	Ha-Shomer ha-Tza'ir
3. Belopolsky Israil	35	Zhitomir	Tiumen'	Dror
4. Blumenkranz Sigizmund	21	Olevsk	Lepinsk	Ha-Shomer ha-Tza'ir
5. Brilliant Ilia	25	Snovsk	Vikulovo	Hitachdut

(continued)

Continued

Name	Age	Previous residence	Place of prison or exile	Party
6. Burak Rachmiel	21		Karakala	
7. Dashnovsky Meer	24	Kremenchug	Temir	Ha-Shomer ha-Tza'ir
8. Epelman Gersh	19	Krivoi Rog	Karo-Kalinsk	Ha-Shomer ha-Tza'ir
9. Farber Moisei	24	Kiev	Bratsk	Hitachdut
10. Gelman Benjamin	22	Kremenchug		Hechalutz
11. Ginsburg Boris	22	Kremenchug	Tashkent	Ha-Shomer ha-Tza'ir
12. Gleiberson Seifert Alexander	26	Minsk	Ust'-Kulom	Hitachdut
13. Yaroslavsky Aron	24	Berdichev	Aulie-Ata	Hitachdut
14. Karasik Mordechai	24		Karo-Kalinsk	Social Zionist
15. Krupnik Arie (Leibl)	21	Olevsk	*etap* [in transit]	Ha-Shomer ha-Tza'ir
16. Levitan David	27	Kiev	Kargopolsko	Hitachdut
17. Madorsky Israil	20	Gomel	Sverdlovsk	Ha-Shomer ha-Tza'ir
18. Ninburg Isaak	24	Melitopol	Inkino	Ha-Shomer ha-Tza'ir
wife Rodoskaja Fania	22	Melitopol	Inkino	Ha-Shomer ha-Tza'ir
child	1	Melitopol	Inkino	Ha-Shomer ha-Tza'ir
19. Novikov Ioseph	30		Minusinsk	Social Zionist
wife Reva	28		Minusinsk	
son Meir	4		Minusinsk	
20. Perelson Yakov	30		Turtkul'	
wife Miriam	24		Turtkul'	
21. Rabinovitch Lev	23	Moscow	Shadrinsk	Hitachdut
22. Rabinovitch Yakov	23	Baku	Khojent	
23. Ratner Juli	31	Leningrad	V. Ural'sk	Social Zionist
wife Sara	32	Leningrad	V. Ural'sk	
son Grigorii	1	Leningrad	V. Ural'sk	
24. Ravsky David	29	Odessa	Djambetta	Young Men's Zionist Social Assoc., EVOSM
25. Rodov Salmon	24	Puchovitshi	Rutsk	Hitachdut
26. Rogovoi Chaim	34	Tarastsha	Astrakhan	Social Zionist
27. Salganik David	34	Kiev	Narym	Social Zionist
28. Saslavsky Aron	24	Kiev	Gusar-Kishlak	Ha-Shomer ha-Tza'ir
29. Serebrani Liova	25	Gomel	Borisovo	Ha-Shomer ha-Tza'ir
30. Smeliansky Girsh	23	Kremenchug	Beli-Skas	Social Zionist
31. Teper Eliakim	24	Balta	Ust'-Sysol'sk	Hitachdut
32. Tahijik Shlioma	24	Odessa	Eniseisk	
33. Zurotschka Ioseph	20	Exaterinoslav	Novo-Urgench	Ha-Shomer ha-Tza'ir

Women

1. Bik Dina	24	Belaja Tserkov'	Atbasar	Social Zionist
son Ioseph	4	Belaja Tserkov'	Atbasar	
2. Birenberg Lia	22	Nikolaev	Samarovo	Ha-Shomer ha-Tza'ir
3. Bolshan Genia	35	Tel Chai	Beresnegovatoe	Hechalutz
daughter Sarra	13	Tel Chai	Beresnegovatoe	
son Yakov	11	Tel Chai	Beresnegovatoe	
4. Freidina Fania	18	Gomel	Kustanai	Ha-Shomer ha-Tza'ir

Continued

Name	Age	Previous residence	Place of prison or exile	Party
5. Galperin Matilda	24	St. Konstantinov Sosnitsa	Bogutshanskaja Sainka	Hitachdut
6. Gushawskaja Galia		Sosnitsa	Solomacha	Social Zionist
7. Kravtshik Menia	22	Poltava	Shirovad	Hitachdut
8. Litvak Bela	22	Rostov	Voronezh	Ha-Shomer ha-Tza'ir
9. Roitman Rachil	25		Kokchetav	Social Zionist
10. Nosovaskaja Maja	25	Kharkov	Borisovo	Ha-Shomer ha-Tza'ir
11. Khinkar Miriam	24	Kremenchug	Atbasar	Social Zionist
12. Shneerson Shtern Hasia	22	Romny	Leningrad	Hitachdut
13. Skullskaja Basia	24	Odessa	Vikulovo	Ha-Shomer ha-Tza'ir

■■■■

21 Draft report of Interdepartmental Committee, Government of Palestine on immigration from Russia[92]

ISA 11/1174/2335, IMM/7/8
October 1930

The Interdepartmental Committee consisting of:
The Chief Immigration Officer
the Officer in Charge of the Criminal Investigation Department
and Mr. Max Nurock, Assistant Secretary

[A]ppointed by His Excellency the High Commissioner to make recommendations for the control of immigration from Russia into Palestine after having given consideration to the procedure at present in force has come to the conclusion that it is possible to reduce the delays and correspondence inherent in the present system without rendering the control of entry in Palestine less effective and at the same time to obtain some security which is not at present available for the cost of repatriating immigrants from Russia whose return to their country of origin in desired.

2 Originally the conditions that governed admission into Palestine from Russia did not differ from those for other countries. It was found however even before the rupture of diplomatic relations with the Union of Soviet Republics in June 1927, that for Police purposes a more careful control over immigrants and travellers from the territories of the Union was desirable. With this rupture of relations the British Consular representatives were withdrawn from Russia and it was then no longer possible to obtain British visas there. There was no desire to exclude from Palestine immigrants in general merely on account of their country of origin and the special and urgent representations made by the Zionist Executive

on behalf of a relatively large class of Jews in Russia on the ground that they were suffering severe persecution on account of their Zionist views were not without effect. To meet the difficulty the concession was granted of accepting a visa for Palestine obtained not, as required elsewhere in the country of domicile, but in a neighbouring country such as Poland or Turkey or even at the Palestine frontier. In practice immigrants from Russia without exception were granted visas on arrival at Jaffa. By this measure the slight safeguard afforded by the local knowledge – in a few cases – of the British Consul was removed.

3 It was found however, almost at once that this procedure made the control far too loose if the security of the country were to be considered and it was decided to tighten it. Within a few months, in October 1927, a revised procedure which involved the reference to Scotland Yard of applications for visas for residents in Russia, except young children and elderly people, and to the CID [Criminal Investigation Department] in Jerusalem in all cases was adopted. It was laid down at the same time that applications for such visas should be accompanied by photographs of the respective recipients so that the Immigration Officer at Jaffa might be able to compare the features of the immigrant or traveller with the photograph of the person on whose behalf the application was made. (See Secret Despatch to Secretary of State of 5.10.27 in Chief Secretary's 16346/27 of 10.10.27). Photographs were, however, available in very few if any cases and this requirement soon became a dead letter. The other requirements however, continued in force. Somewhat later a new condition, the requirement of photographs on arrival, so that the immigrants could later be identified if they came under unfavourable notice by the Police was instituted. At the same time the Zionist Executive or any other applicant in Palestine for a traveller or immigrant from Russia was called upon to furnish a certificate to the effect that the prospective immigrant or traveller was not connected with the Communist party, the left Wing of the Po'alci Tziyon or any kindred movement. (Chief Secretary's 2135/29 of 12.3.29 and 9.4.29). The Zionist Executive, however, objected at once to this last condition and it was modified to the effect that the applicant should inform the Chief Immigration Officer of the political affiliations of the prospective immigrant as far as they were known (Chief Secretary's 2135/29 of 4.6.29). From that date in almost all if not all of these cases the certificate has been to the effect that nothing is known of the political affiliations of the prospective immigrant.

4 With the reappointment of British Consuls in the territories of the Soviet Republics the practice of granting visas at the Palestinian port has been brought to an end and immigrants and travellers from Russia to Palestine are now required to obtain their visas at a British Consulate in Russia. Otherwise the procedure remains unchanged. It is in brief as follows:

The application, if its grant does not conflict with the Immigration Ordinance Regulations is referred to Scotland Yard (except in the case of young children and elderly persons) and to the Deputy Commandant CID in Jerusalem. If neither knows of any objection either to the prospective immigrant or to his referees, an Immigration Certificate is issued and a British Consul in Russia is authorized to grant the required visa provided the applicant holds a valid national passport.

This holder however is not permitted to enter Palestine until he has been photographed. The photographs are placed at the disposal of the CID.

5 The following defects have disclosed themselves in this practice: The prospective immigrants in almost every case are unknown both to Scotland Yard and the Palestine CID or if known adopt names for their immediate purpose with which neither is acquainted. As a consequence Scotland Yard has never objected to any suggested immigrant and the Palestine CID also has very seldom if ever had any good reason to object. When the visas were granted at Jaffa there was no further safeguard. Now that they are granted at Moscow or Leningrad there is still very little if any safeguard, for British Consuls have no means of learning anything of applicants apart from very exceptional cases. As a result practically nothing is known of any of these people when they arrive in Palestine and when it is later found advisable to deport any of them they are as rule without means and the cost of deportation falls on the Palestine Treasury. Reference to Scotland Yard at any rate serves no useful practice. It, however, involves delay in deciding whether or not the visa is to be authorized and correspondence between the Immigration and Travel Section and the Chief Secretary, the High Commissioner and the Secretary of State, the Secretary of State and the Home Office and possibly the Home Office and Scotland Yard.

6 In view of the foregoing the Committee makes the following recommendations:

(a) Travellers and Immigrants from Russia shall pass through an identical procedure and in view of the difficulty, even the impossibility, in the case of travellers from Russia whose political bona fides is unquestionable, of returning to that country, no visa for a traveller shall be authorized unless the applicant, in the event of desiring to settle in Palestine, is able to qualify for the receipt of an Immigration Certificate.

(b) No visa should be authorized in favour of any person who cannot furnish a reference in Palestine, to whom the CID has no objections, who has furnished a certificate in the following form, 'I. A.B. certify to the best of my knowledge and belief that Y is of good character and politically "desirable" ' and who is prepared to deposit with the Chief Immigration Officer the sum of £.20 in respect of every immigrant or traveller on whose behalf he makes application, this amount to be returned after the expiration of two years from the entry into Palestine or on the departure of the immigrant or traveller from Palestine whichever date be the earlier, unless the sum or any portion of it has been used in defraying, especially in the case of deportees, the cost of the immigrants' maintenance in Palestine and return to Russia.[93] (In the case of responsible institutions, such as the Jewish Agency and the Hebrew University, deposits would not be required, but undertakings to pay the cost of maintenance and repatriation over a period of three years would be accepted instead.)

(c) The return to Palestine of a former resident who had left Palestine for Russia would not as a rule be authorized, but exceptions would be made when such a course is considered justified by the Deputy Commandant, CID, and the Chief Immigration Officer.[94]

(d) The Secretary of State should be asked to make representations to the Foreign office to the effect that a British Consulate be opened at Odessa at as early a date as is practicable. (Practically all immigrants and travellers to Palestine from Russia have to pass through Odessa where they could obtain their visas when on their journey. The present practice of granting visas only at Moscow and Leningrad imposes appreciable hardship, expense and loss of time on the immigrant or traveller with no compensating advantage.)

(e) In order to introduce the requirement of a deposit in cash as proposed in (b) above the necessary amendment should be made to the Immigration Ordinance as recommended by the Chief Immigration Officer in his IMM/1/1/A of 30.4.29 to the Chief Secretary.

Chief Immigration Officer
Deputy Commandant CID
Assistant Secretary

■ ■ ■ ■

22 Y. Gruenbaum, Jewish agency to Director of Immigration, Government of Palestine

ISA 11/1174. IMM/7/6 II
Jerusalem, 26 February 1934

The Director
Department of Immigration
Jerusalem.

Sir,

I have the honour to approach you on behalf of a number of political refugees who are persecuted in Russia for their Zionist activities and whose names are given in the enclosed list.[95] We are informed that the Soviet authorities have agreed to permit them to proceed to Palestine in lieu of continuing their exile and imprisonment, but that they would be able to avail themselves of this 'exchange' only in case the authority for the admission into Palestine is given at once. Otherwise, there is the serious risk of their being sent again to exile and imprisonment, and perhaps losing even the chance of ever coming to Palestine.

I would therefore ask that you be so good as to look into these cases and issue the desired authority immediately, the matter being of outmost urgency.

To avoid any possible delay, we beg to state that we agree to undertake responsibility to defray the cost of their return to Russia, should such necessity arise, within three years from the date of their entry to Palestine, provided that our liability does not exceed LP.10. – in respect of each. The Executive reserve themselves the right to revert to these particular cases after the close of the negotiations regarding the general issue which was raised in our letter of February 7th, 1934.

It will be greatly appreciated if the necessary authority in respect of these refugees would be sent by telegram, the cost of which will be paid by us.

I have the honour to be,
Sir,
Your obedient servant,

Y. Gruenbaum
Executive of the Jewish Agency

Notes

1 An additional letter by Unshlikht of the same date, certified that 'the office located at Kuznetskii Most St. 16, currently occupied by the former Moscow Political Red Cross, with all its contents, is transferred to E.P. Peshkova for work in rendering assistance to political prisoners and their families', GARF f. 8409, d. 11, l. 3.
2 This document dated 10 February 1923 was located in another file, f. 445, op. 1, d. 167, ll. 185, 206–7. The CC CPU opposed legalization of Hechalutz.
3 A major administrative-territorial division; sometimes referred to in translation as a province.
4 In the course of their armed struggle against Bolshevik power, these nationalist forces engaged in widespread pogroms against the Jewish population.
5 The émigré group *Smena Vekh* (Changing Landmarks) advocated in the early 1920s acceptance of the Bolshevik revolution by the intelligentsia, in order to transform the regime from within and safeguard the Russian national state.
6 Professor Boris Brutskus, an agronomist, was among many academic and professional activists arrested in late summer 1922; only his wife's plea to the GPU offices in Petrograd prevented his deportation from Soviet Russia. He may have been the brother of Dr Julius Brutskus, a leader of the Zionist Organization in Russia both before and after the revolution of 1917.
7 Established by Baron de Hirsch in 1891 to facilitate mass emigration of Jews from Russia to agricultural colonies particularly in Argentina and Brazil.
8 Petr Nikolaevich Wrangel was Commander-in-Chief of all White armies operating from the Crimea. He established a provisional government there, but in 1920 the Red Army broke through his defences and he fled to Turkey. His defeat ended White Russian resistance to the Bolsheviks.
9 In fact, it was the NKVD, not Narkomindel, and the date was 30/VIII.
10 Vserossiiskii tsentral'nyi ispolnitel'nyi komitet, All-Russian Executive Committee of Soviets.
11 The telegram was sent by the Central Bureau to all regional *evsektsii* on 4 September 1922.
12 *Guberniia* committee secretaries, in this case of the Jewish Sections (*evsektsii*) attached to the regional RCP committees.
13 Agricultural farms; usually cooperative or collective.
14 A Yiddish term for General Zionists, that is, members of the Organization of Russian Zionists.
15 The Zionist–Socialist Party held its all-Russian Fourth Conference in early 1924, but it took place in Leningrad, not Moscow.
16 Eliezer (Leizer) Zusmanovich Shub, b. 1903; like many of the arrested Zionists whose names are listed in this report, was granted substitution and arrived in Palestine on 14 July 1924 on the Soviet steamer *Novorossiisk*.

17 The reference here is to ESSM – the Jewish Socialist Youth League.
18 *Sotsialisticheskii vestnik* was the central print organ of the Russian Social Democratic Labour Party (better known as the Mensheviks). It was founded by L. Martov in 1921 after his emigration from Soviet Russia and published up to 1940 in Berlin and Paris and subsequently in New York.
19 Ginda Borukhovna Malkova, was sentenced to three years in *politizoliator* (1924–1927) and three more years of exile (1927–1930). Her requests for substitution were denied at least twice (1930, 1931). While in exile, she married Zerubavel Evzerikhin, a leader of Legal Hechalutz, and they had a daughter. A physician by profession, Malkova died in 1935 while fighting cholera in Turukhansk in Eastern Siberia.
20 Solomon D. Shpiro was sentenced in May 1924 to three years of exile with the option of leaving for Palestine; A passenger named Shlomo Shapira – presumably the same individual – arrived on the *Novorossiisk*.
21 Shneur Aronov, a leader of Legal Hechalutz, was sentenced in May 1924 to three years of exile with the option of leaving for Palestine; arrived on the *Novorossiisk*.
22 Dan Pines, a leader of Legal Hechalutz.
23 The term used in the original is *intelligent* – a social category denoting a certain level of higher education.
24 A state institution that carried out the purchase of goods from private producers.
25 Nachum Plotkin was sentenced in May 1924 to three years of exile with the option of leaving for Palestine; arrived on the *Novorossiisk* with his wife and two children.
26 A monthly literary and scholarly journal. It was published in Moscow from 1921 through 1942.
27 This reference is probably to Yakov Iegudovich Iavno, b. 1899, who is known to have been arrested on 14 March 1924. He was sentenced to two years of exile but allowed to leave for Palestine on the *Novorossiisk*.
28 Samuil Gurvich left for Palestine on the *Novorossiisk*.
29 Iosef Mendelson, b. 1896, arrived in Palestine on the *Novorossiisk* 14 July 1924.
30 Binyamin Vest (B. West), b. 1896, arrived in Palestine on the *Novorossiisk*. He was one of the founders of the STP and Illegal Hechalutz.
31 Isai (Se'adia) Lazarevich Gol'dberg, b. 1884, was sentenced to three years of exile but received substitution and left on the *Novorossiisk* with his wife and son.
32 In spite of the discrepancy in the first initials, the reference is almost certainly to Professor Grigorii (Tzvi or Girsh) Belkovskii, b. 1865; sentenced to two years of exile but received substitution; he arrived in Palestine on 2 September 1924 on the Italian steamer *Milano* from Istanbul.
33 Tsentral'noe Upravlenie Legkoi Promyshlennosti – the Central Administration for Light Industry.
34 Roza Gel'chinskaia arrived in Palestine on the *Novorossiisk*.
35 Moskovskii sovet narodnogo khoziaistva – the Moscow Council of National Economy; a local organ for the administration of industry. It existed from 1917 through 1932.
36 Genrikh Grigor'evich Iagoda (1891–1938) was Deputy Chief of the GPU/OGPU from 1923 to 1929. From 1924 to 1934 he was a member of the Special Commission of the Soviet security police.
37 T.D. Deribas. The Secret Department was responsible for collecting information on political parties and organizations.
38 Viacheslav Rudol'fovich Menzhinskii (1874–1934) was Deputy Chief of the Cheka/GPU/OGPU from 1923 to 1926. He served as Chief of the OGPU, 1926–1934.
39 Dzerzhinskii is referring here to the debates within the Bolshevik Party about its national policy: one wing – the 'assimilators' – advocated the assimilation of Russia's nationalities to create a unified workers' movement.
40 See V.I. Lenin, 'Detskaia bolezn' "levizny" v kommunizme' in V.I. Lenin, *Sobranie sochinenii*, Vol. 31 (Moscow: Gospolitizdat, 1956), pp. 1–97, http://www.markists.org/archive/lenin/works/1920/lwc/index.htm.

41 Dzerzhinskii apparently misused the term *poiti navstrechu* here. Literally 'to reach out', he uses it to denote 'cracking down on'.
42 The so-called Schedule, or Labour Schedule, was issued by the Government of Palestine twice annually (in October and April) and stipulated the number of Immigration Certificates to Palestine to be issued to Jewish 'labourers' over the next six months. Responsibility for the distribution of these certificates, for choosing individual immigrants, and for guaranteeing their solvency rested with the PZE.
43 Mordechai Nemirovskii (Namir), b. 1897, a member of TzS, arrived in Palestine on the *Chicherin* on 16 December 1924. In later years he held important positions in the Histadrut and served as a Knesset member, ambassador to the Soviet Union, Minister of Labour, and mayor of Tel Aviv.
44 Lieutenant-Colonel F.K. Kisch, Chairman of the PZE from 1922 to 1931.
45 The text reads *Ch. klali* – presumably *Chalutz klali*, that is, Hechalutz.
46 Also called the Eretz Yisrael Offices, these were set up by the Immigration Centre of the PZE in countries throughout Central and Eastern Europe to select immigrants and help arrange their passage.
47 That is, the Immigration Centre (Merkaz ha-Aliya).
48 Reference is to the broadsheet distributed by TzS and its youth movement, TzS Yugend on 16 August 1924 and the arrests that followed on 2 September 1924. See the Introduction to this volume.
49 The intention is clearly to the Procurator.
50 An apparent mistranslation by the author of the Hebrew/Russian *angina*, that is, strep throat.
51 Attached to the letter was a list of names of 29 men and 8 women.
52 Yakov Genkin. The 4th Section (*otdelenie*) was entrusted with collecting information on Zionists.
53 Albert Montefiore Hyamson, a long standing British Zionist, joined the Government of Palestine with the help of the Zionist Executive in London, serving from 1921 to 1934 as Chief Immigration Officer.
54 According to documents preserved in the same file, the PZE forwarded this letter to the Zionist Executive in London, adding that it had already provided multiple certificates for Zionist prisoners (168 between September 1926 and April 1927), that financial aid was beyond its means, and that it believed it 'possible to awaken known individuals in Russia to come to the aid of these prisoners', CZA S25/f. 2424.
55 Owing to the severance of diplomatic relations between Great Britain and the Soviet Union.
56 Memorandum.
57 Max Nurock, Personal Secretary to High Commissioner Herbert Samuel and later Assistant Chief Secretary of the Government of Palestine. He was one of several British Zionists serving in high government positions in Palestine. With the establishment of the State of Israel he joined its diplomatic service and was Ambassador to Australia.
58 See Document 12: Memorandum of the Norwegian Legation on the procedure for obtaining Palestine visas for Soviet citizens.
59 Field-Marshal Herbert Onslow Plumer was High Commissioner for Palestine (1925–1928).
60 Leopold Amery served as Britain's Secretary of State for the Colonies.
61 Lieutenant-Colonel Sir George Stewart Symes served as Chief Secretary to the Government of Palestine during Plumer's term as High Commissioner (1925–1928). His telegram referenced by Plumer could not be found in the archives.
62 The dispatch referred to could not be located in the archive.
63 The Secretary approved the new procedures in his secret dispatch of 29 December 1927 (referred to by Plumer in his secret dispatch to the Secretary for the Colonies 26 July 1928, PRO, CO 733/159/16). A letter from the Chief Secretary marked 'confidential' and dated 27 December 1927 informed the Chief of Police to refer all

lists received by the PZE from the Committee to Aid Political Prisoners to the Foreign Office and Scotland Yard. ISA, ISA 11/1174, IMM/7 I.
64 Yakov Solomonovich Arav, known also by the name Foma L'vovich, a member of the TzS was arrested in Vinnitsa in April 1926 and arrived in Palestine on the ship *Tobol'sk* on 24 May 1927.
65 Dom predvaritel'nogo rassledovaniia – the House of Preliminary Investigation – was where those arrested and convicted were temporarily detained before sending them to the places where they would serve out their sentences.
66 Lota Borisovna Sandomirskaia headed the Kharkov office of Pompolit. See Chapter 2: Out of the Soviet Union.
67 The letter was apparently written around 6–7 February and dated according to the day of its registration in Pompolit.
68 The official name used between 1921 and 1924 for Ashkhabad, the capital of the Turkmen SSR.
69 In the original, PP – *polnomochnyi predstavitel'*.
70 Political exiles were required to appear and register at the local office of the security services (GPU, OGPU) at regular intervals – weekly, daily or twice-daily. The frequency was often subject to the whims of local officials.
71 This refers to the procedure that existed up to this point under which the form was filled out after receipt of the entry certificates and was presented to the immigration authority upon entrance to Palestine.
72 Zelik Kunin; see GARF, f. 8409, op. 1, d. 176, ll. 252–3.
73 On the Special Commission see Chapter 2: Out of the Soviet Union.
74 ZAGS (*Zapis' aktov grazhdanskogo sostoianiia*) – Division for the Registration of Civic Acts – a state institution where births, marriages, divorces and deaths were registered.
75 The letter was actually dated 27 April.
76 To Bender, Peshkova wrote:

> ...in response to your message, I can report that in order for your wife to leave for Palestine on the same visa as you, it is necessary to send her the certificate you received and the marriage license from ZAGS to be presented to the Mission and so that the necessary notations can be made. They promise to enter the visa into your wife's passport if you present the certificate with the marriage license before your departure, but in Palestine, you would have to make the necessary declaration so that when you wife enters there will be no obstacles.
> GARF, f. 8409, op. 1, d. 179, l. 196

77 Administrativnyi otdel Moskovskogo Soveta – The Administrative Council of the Moscow Soviet.
78 The letter bears a handwritten notation in pencil: 'Summon Molodetskaia. Inform her of her husband's response. Vinaver.'
79 Arav was clearly referring here to the Immigration Department.
80 The arrangements Arav discusses in this letter reflect the changes brought about by the break in diplomatic relations between Great Britain and the Soviet Union and the departure of the British Mission from Moscow.
81 A cooperative settlement founded in 1921 in the western Jezreel Valley.
82 British policy on visas for spouses changed over time. In the early phases of substitution, wives of men who got married after receiving their Immigration Certificates were immediately granted visas. This latitude was often exploited by the Zionists to increase the number of certificates available, or to gain entrance to Palestine for women not otherwise eligible. Many men entered into fictitious marriages, which were annulled soon after they arrived in Palestine. Aware of this abuse, the British began to place restrictions on spousal visas and matters were further complicated in cases where the husband was forced to leave for Palestine before his wife could secure a visa, usually

because passport, certificate or visa was about to expire. The case of Gushanskii, a member of Ha-Shomer ha-Tza'ir, was especially tragic: After three years in exile, he arrived in Palestine in November 1928; in August 1929 he was accidentally killed while working on the construction of the hydroelectric dam in Naharayim. His wife, Maria Fedorovna Lopatnikova-Gushanskii, arrived in Palestine just two months before his death. A non-Jew, she returned a few years later to the Soviet Union.

83 The original is in German.
84 The reference is to Hechalutz (Hebrew for 'pioneer').
85 Avrum Merezhin; see Document 2: Memorandum of A. Merezhin, Central Bureau of the Evsektsiia to the Politburo.
86 In fact, administrative sentencing was applied to all those arrested by the GPU for political crimes, regardless of party affiliation.
87 Frida Eventov was 23 at the time of her death. She had been a member of Illegal Hechalutz as well as EVOSM. See Ar'ye Refa'eli, *Ba-Ma'avak li-Ge'ula. Sefer ha-Tziyonut ha-Rusit mi-Mahapekhat 1917 ad Yameinu* (Tel Aviv: Davar and Ayanot, 1957), p. 196.
88 The murdered prisoner was Samuil Bronshtein. A circular sent by Ha-Shomer ha-Tza'ir on 15 April 1928 explained that the shooting took place after he demanded that the guards stop cursing the prisoners: Avraham Itai, *Korot Ha-Shomer ha-Tza'ir be-S.S.S.R.; No'ar Tzofi Chalutzi – NeTzaCh* (Jerusalem: Ha-Aguda le-Cheker Tfutzot Israel, 1981), p. 334.
89 Avram Shkipka, 22 years old, was also a member of EVOSM. His death occurred on 10 June 1927. Refa'eli, *Ba-Ma'avak li-Ge'ula*, p. 198.
90 That is, of the present month.
91 A hand written note says, 'Received 4/July/29'.
92 The draft was sent by Chief Immigration Officer Hyamson to several officials for their comments (24 October). The final report, dated 28 October, was more formal and crisp in style, and omitted all reference to the objections of the Zionist Executive or its role in developing the policy. Instead, it placed its recommendations in the context of immigration policies in other British colonies. Two minor substantive differences between the draft and the final report are noted.
93 The final report called for the deposit to be left with the Chief Immigration Officer for three years.
94 The language of the final report was less restrictive, stating only that 'Applications by or on behalf of former residents of Palestine for permission to return to Palestine should be considered each on its merits.'
95 Not attached to the original archival document.

Glossary

aliya The Hebrew term for Jewish immigration to the Land of Israel, literally – 'ascent'
CC Central Committee
chalutzim Literally, pioneers or vanguard; Zionist ideology stressed the need for the Jewish people to normalize their lives by changing their economic structure and to become workers and farmers, who would settle in the Land of Israel and work the land
CID Criminal Investigation Department, Government of Palestine
Comintern Communist International, world organization of communist parties; existed 1919–1943
CPU Communist Party of Ukraine
CZA Central Zionist Archives
Eretz Yisrael The Land of Israel – the term used by Jews for pre-state Palestine
ESSM *see* Jewish Socialist Youth League
EVOSM *see* United All-Russian Organization of Zionist Youth
Evsektsiia (pl. *evsektsii*) Jewish section of RCP(b) Central Committee, founded 1918; operated alongside additional national sections – German, Polish, Yugoslav, Lithuanian, Estonian, Czech and Hungarian; abolished in 1930
GARF Gosudarstvennyi arkhiv Rossiiskoi Federatsii (State Archive of the Russian Federation)
GPU Gosudarstvennoe politicheskoe upravlenie (State Political Directorate), security police; became OGPU in 1923
Ha-Shomer ha-Tza'ir Zionist socialist youth movement, founded in Europe in 1916, to prepare Jewish youth for *kibbutz* life in Israel
Ha-Shomer ha-Tza'ir ha-Amlani Childrens' and youth movement, organized beginning in 1924 by members of EVOSM; the Toilers' Wing of Ha-Shomer ha-Tza'ir in Soviet Russia; affiliated with EVOSM
Ha-Shomer ha-Tza'ir ha-Ma'amadi Childrens' and youth movement organized by Jewish students in 1922; combined Zionist and socialist teachings; later known as the Soviet wing of Ha-Shomer ha-Tza'ir; it broke with the world organization in 1930
Hechalutz (pioneer/vanguard) World Zionist organization begun in Russia to promote training primarily for agricultural work in Palestine; divided in

Russia into two branches: Legal and Illegal or National Hechalutz – the latter made up of those opposed to registration of the organization with the Soviet authorities

Histadrut   General Federation of Labour in Eretz Yisrael established in 1920

Hitachdut (federation)   A world Zionist labour movement formed in 1923 through the union of Tze'irei Tziyon and Ha-Po'el ha-Tza'ir; used also interchangeably with STP – Zionist Toilers' Party, which affiliated itself with it

Immigration Centre of the General Federation of Labour (Histadrut)   Merkaz ha-Aliya, often referred to simply as the Immigration Centre; translated into Russian as Tsentral'nyi otdel rabochei immigratsii

ISA   Israel State Archives, Jerusalem

JDC   Joint Distribution Committee

Jewish Socialist Youth League   ESSM, Evreiskii Sotsialisticheskii Soiuz Molodezhi

Komzet   Komitet po zemel'nomu ustroistvu trudiashchikhsia evreev, Committee for Agricultural Settlement of Toiling Jews

Mapai   Mifleget Po'alei Eretz Yisrael, Party of Workers of Eretz Yisrael

Merkaz ha-Aliya   *see* Immigration Centre

OGPU   United State Political Directorate *see* GPU

Orgburo   Organizational Bureau of the Central Committee of the RCP

OSO   Osoboe soveshchanie OGPU, Special Commission of the Soviet security police; given extra-judicial power to deport, exile and imprison in a concentration camp or a political prison, any individual implicated in anti-Soviet activity

Ozet   Vsesoiuznoe obshchestvo po zemel'nomu ustroistvu trudiashchikhsia evreev v SSSR, All-Union Society for Agricultural Settlement of Toiling Jews in the USSR

Po'alei Tziyon (Workers of Zion)   Social Democratic Zionist Party founded in Russia in the early 1900s by Ber Borochov; in Soviet Russia it refashioned itself into the Jewish Communist Party (EKP-Po'alei Tziyon). A World League of Po'alei Tziyon was founded in 1907; after its split in 1920 the right wing affiliated with the World Zionist Organization while Left Po'alei Tziyon rejected the WZO and sought admission into the Communist International

Pompolit   Pomoshch' politicheskim zakliuchennym, Committee to Aid Political Prisoners – also known as Aid to Political Prisoners and as the Political Red Cross

PZE   Palestine Zionist Executive set up in 1921 by the World Zionist Organization; replaced in 1929 by the Jewish Agency

RCP(b)   Russian Communist Party (Bolshevik)

RGASPI (formerly RTsKhiDNI)   Rossiiskii gosudarstvennyi arkhiv sovremennoi politicheskoi istorii – Russian State Archive for Contemporary Political History

RSDLP   Russian Social Democratic Labour Party; in the period covered by this volume the official name of the Mensheviks

Sovnarkom (SNK) Sovet narodnykh komissarov – Council of People's Commissars

SRs Members of the Socialist Revolutionary party

STP (STP TzTz) Zionist Toilers' Party; abbreviation is from the Russian Sionistskaia Trudovaia Partiia; known in Hebrew as Tze'irei Tziyon-Hitachdut

Tze'irei Tziyon (TzTz) A fraction of the Organization of Russian Zionists established in 1913 primarily by students with demands for greater democracy within the movement and a greater emphasis on practical measures to prepare pioneers for immigration to Palestine. In 1920 the movement split into the Zionist–Socialist Party and the Zionist Toilers' Party

TzS (TzS TzTz) Zionist–Socialist Party; abbreviation is from the Yiddish

TzS Yugend/TzS Yugend Farband Zionist Socialist Youth League; youth movement of the TzS

United All-Russian Organization of Zionist Youth Edinaia Vserossiiskaia Organizatsiia Sionistskoi Molodezhi (EVOSM), sometimes referred to as simply 'Histadrut' (the Organization) or Histadrut Ha-No'ar ha-Tziyoni

VTsIK All-Russian Executive Committee of Soviets (Vserossiiskii tsentral'nyi ispolnitel'nyi komitet)

Zionist Executive Executive of the World Zionist Organization in London, known in Hebrew as Ha-Hanhala ha-Tzionit

Index

Abrabanel, Avnir 78, 83
Achdut ha-Avoda 43, 81
Afikim, Kibbutz 81
Agreement on Repatriation (1920) 38
agricultural training farms 3, 5, 116;
  see also Joint Distribution Committee;
  Komzet; Mishmar; Ozet; Tel Chai
Aid to Political Prisoners see Pompolit
Aikhenval'd, Iulii 22
Aktiubinsk 113
Alexander II 2, 24
Algemeyn Tziyen see General Zionists
All-Russian Committee to Aid the
  Starving 21
All-Russian Executive Committee of
  Soviets (VTsIK) 7, 13, 14, 88
All-Russian Extraordinary Commission
  for Fighting Counter-revolution and
  Sabotage (VChK, Cheka) 6, 12, 27,
  36, 130
All-Russian Zionist Conference 14
All-Union Society for Agricultural
  Settlement of Toiling Jews (Ozet) 13
Amery, L.C.M.G. 100, 131
antisemitism 2, 86, 116, 129; of
  nationalist groups 86–7, 129; see also
  pogroms
Arabs in Palestine 46, 52, 55, 61, 78
Aran (Aronovich), Zalman 81
Arav, Yakov (Foma L'vovich) 102–11, 132
Arest, Avraham 81
Argentina 114
Argov (Grabovskii), Me'ir 81
Arkhangel'sk 26
Arlosoroff, Chaim 52
Aronoff, Shne'ur 73
Ashkhabad see Poltoratsk
Australia 114
Austria 114

Balfour Declaration 2, 50, 55, 58
Barlas, Chaim 73
Bar-Yehuda (Idelson), Israel 81
Belgium 114
Belkovskii, Tzvi 56, 130
Belorussia 3, 4, 8, 15, 19, 86, 120;
  nationalist organizations in 87
Ben-Gurion, David 43, 48, 54, 65, 69, 71;
  in Moscow 11, 33, 56, 66
Bentwitch, Norman 57, 58, 72
Berdiaev, Nikolai 22
Bessarabia 114
Bialik, Chaim Nachman 67
Bogdanovich, Iurii 24
Bogdanovsky, M. 47
Bokii, Gleb 17
Borochov, Ber 73
Borovaia, Bella 75
Borovaia, Raisa 75
Brandeis, Louis 64, 74
Brazil 114
Breinin, Chaim 70
British Mandatory Government of
  Palestine 33, 34, 41–4, 53, 55–9; Chief
  Immigration Officer 96, 97, 99, 101,
  102, 122, 123, 125, 126, 127, 128;
  immigration policies 34, 39, 41, 58, 73,
  55, 122, 125, 126, 132; political
  screening of immigrants 50, 59, 61, 62,
  65, 73, 101, 126; see also visas for
  Palestine
Brutskus, Boris 87, 129
Brutskus, Julius 129
Bucovina 114
Bulgaria 114
Bund 87
Bureau for Administrative Deportation of
  the Intelligentsia 21
Butyrskaia Prison 14, 18

Canada 114
Caucasus 80
Central Committee to Aid Russian Zionists 51
Chaikin, Bat Sheva 81
Chalamish, Aviva 70
Chancellor, John 63, 74
Chardzhui 104
Chasidim 80
Cheka *see* All-Russian Extraordinary Commission for Fighting Counter-revolution and Sabotage
Cheliabinsk 17
Cherkasskaia, Leah 18
Chicherin 40, 59, 67, 72
Chief Immigration Officer *see* British Mandatory Government of Palestine
Chile 114
Chimkent 111, 120
Chuvash region 68
Circle to Aid Prisoners and Exiles (Kruzhok pomoshchi katorge i ssylke) 26
Civil War 2, 3, 6, 21, 25
Clerk, George 102
Cohen, Dennis 72
Comintern *see* Communist International
Committee for Agricultural Settlement of Toiling Jews (Komzet) 7, 9, 13, 117
Committee for Aid to Political Prisoners in Russia 25
Committee of Assistance to Children 26
Committee to Aid Political Prisoners *see* Pompolit
Committee to Help Persecuted Russian Zionists 74
Communist International (Comintern) 59, 61, 63, 134
Communist Party of Ukraine 86
Communist Youth League (Komsomol) 7
Constantinople 35, 60, 62, 99, 101, 109, 110
Constituent Assembly 2
Constitutional Democrats (Kadets) 66, 87
Council of People's Commissars (Sovnarkom) 7, 13, 14, 103
Crimea 3, 9, 15, 19, 25, 45, 51, 69, 89, 116
Crimea Project 52
Criminal Investigation Department (CID) 126, 127, 128
Czechoslovakia 115

Dan, Fedor 21
Davar 43, 46, 62, 67
de Hirsch, Maurice 129

deportation 7, 17, 21–2; of intellectuals 10, 21, 22; of Zionists *see* substitution emigration
Deribas, Terentii Dmitrievich 15, 23, 91, 95, 130
Dobkin, Eliyahu 67
Dobkin, Levi 47, 68
Driker, Genrikh 75, 83
Dror 6
Dzerzhinskii, Feliks 7, 8, 15, 21, 22, 23, 24, 26, 27, 28, 29, 30, 39, 91, 94, 95, 130; biographical details 27; on persecution of Zionists 91, 94, 96

Egypt 59
Ein Charod 80
Ekaterinburg 14, 36
Ekaterinoslav 120
Emigré Fund (Immigrantskaia Kassa) 26
Eniseisk 78, 120
Eretz Yisrael Offices *see* Palestine Offices
Eshkol (Shkolnik), Levi 56, 71
ESSM *see* Jewish Socialist Youth League
Esterlis, Akiva 78, 84
Estonia 115
Evory, Esmond 52
EVOSM *see* United All-Russian Organization of Zionist Youth
Evsektsiia 6, 7, 9, 67, 83, 86, 88, 91, 116, 117, 121, 134
exile, internal 18, 19–20, 115; conditions of 20, 118, 119, 120; *see also* deportation; minus zones; substitution emigration
Ezhov, Nikolai 30, 39

Federation of Jewish Labour in Palestine *see* Histadrut
Figner, Vera 24, 26, 28
Finland 115
France 115
Frantz Mehring 40

Galicia 115
Galili (Iskoz), Elazar (Lasia) 67, 74
Gal'perin, Boris 23
Gelman, Dr 33
General Federation of Labour in Eretz Yisrael *see* Histadrut
General Zionists 5, 44, 45, 89
Genkin, Yakov M. 15, 23, 131
Germany 115
Gervitz, Bella 84
Gervitz, Mendel 78, 84

Goldberg, Se'adia 44, 54, 67
Gomel 15, 86
Gomel Province 15
Gordon, Yehoshua 68
Gorevoi, Yakov 79, 84
Gorkii 75
Gorky, Maxim 25, 26, 27, 30, 100; work on behalf of political prisoners 25
Gorlan 78
GPU (State Political Directorate) 6, 8, 12, 15, 17, 21, 23, 25, 27, 62, 78, 83, 85, 86, 88, 89, 103, 115, 116, 117, 118, 119, 120, 121; Ukrainian 8; *see also* Cheka; NKVD; OGPU
Great Britain 98, 99; relations with Soviet Union 52, 57, 59, 62, 71; *see also* British Mandatory Government of Palestine; Moscow: British Mission in; Scotland Yard
Great Terror 16, 18, 30, 76
Grinberg, Chaim 74
Gruenbaum, Yitzchak 64, 128, 129

Hadassah Hospital 82
Hagana 81, 82
Haifa 35, 78, 81
hakhshara see agricultural training farms
Halperin, Chaim 34, 47, 49, 51, 68, 73, 111, 112
Halperin, Matilda 78, 83
Halperin (Gal'perin), Yechi'el 83
Halpern, Dr 69, 70
Ha-Po'el ha-Tza'ir 45
Ha-Shomer ha-Tza'ir 5, 9, 12, 16, 19, 43, 67, 74, 75, 79, 80, 81, 92, 116, 120, 123, 124, 125, 134; last meeting of 13
Ha-Shomer ha-Tza'ir ha-Amlani (the Toilers' Wing of The Young Guard) 5, 134
Ha-Shomer ha-Tza'ir ha-Ma'amadi (the Class/Socialist Wing of The Young Guard) 6, 9, 13, 66, 134
Hebrew language and culture 3, 4, 5, 44, 45, 67, 78, 121
Hebrew University 127
Hechalutz 4, 5, 6, 7, 8, 9, 11, 12, 14, 15, 42, 43, 45, 47, 48, 49, 54, 55, 56, 63, 67, 68, 70, 72, 77, 80, 86-8, 90, 92, 93, 97, 116, 117, 118, 120, 124, 134; Illegal (National) 6, 9, 11, 12, 45, 48, 65, 69, 73, 77, 80, 83, 84, 116, 117, 120, 130, 133, 134; Legal 12, 13, 17, 43, 45, 47, 48, 67, 70, 72, 73, 80, 84, 116, 130, 134; World (Berlin) 45, 46, 47, 48, 67, 68, 92-3
Hechaver 14, 19, 67, 68, 81, 89, 90
Henderson, Arthur 52
Histadrut 2, 33, 34-53 *passim*, 56, 63-7 *passim*, 80, 81, 82, 107, 131; agricultural centre 71; third conference (July 1927) 67; *see also* Immigration Centre of the Histadrut
Hitachdut 17, 135; Soviet wing of *see* Zionist Toilers' Party
Horenshtein, David 77, 83
Hungary 115
Hyamson, Albert Montefiore 56, 57, 58, 61, 62, 63, 64, 72, 96, 97, 98, 100, 131

Iagoda, Genrikh 15, 17, 30, 39, 91, 130
Iaroslavl' Prison 17, 18
Idelson, Beba 81
Idelson, Yisrael 66, 68
Il'ich 40
Illegal Hechalutz *see* Hechalutz
Immigration and Travel Department (Immigration Section, Permit Section) of the Government of Palestine 41
Immigration Centre of the Histadrut (Merkaz ha-Aliya) 34, 41, 42, 43, 46-9, 50, 53, 63, 64, 65, 68, 69, 72, 73, 92, 93, 111, 112
Immigration Certificates 33, 41, 48, 49, 52, 53, 59, 60, 61, 63, 66, 70, 73, 77, 78, 80, 97, 98, 101, 102, 110, 122, 123, 126, 127, 131
Immigration Department of the Palestine Zionist Executive (PZE) 41, 49, 51, 53, 54, 55, 58, 63, 64, 65, 74, 96, 109
Immigration Ordinance 58, 72, 97, 122, 126, 128
Intourist 33
Irkutsk 15
Iskoz, El. *see* Galili
Italy 115
Ivanovskaia-Voloshenko 30
Izhma 111

Jacobs, J. 59, 93
Jaffa 22, 34, 35, 39, 59, 60, 62, 102, 105, 107, 108, 126, 127
Jambul 83
JDC *see* Joint Distribution Committee
Jewish Agency 33, 50, 51, 53, 62, 64, 70, 80, 127, 128, 129

Jewish Colonization Association (JCA) 51, 61, 88
Jewish communal institutions (*kehillot*) 3, 11
Jewish Communist Party 73
Jewish Emigration Society 69
Jewish philanthropies 50–3, 67, 87, 88, 115; *see also* Joint Distribution Committee; Zionism in Soviet Russia: funding of
Jewish Socialist Youth League (ESSM) 6, 80, 81, 89, 135
Jews, Soviet policy toward 1, 3, 4; *see also* Evsektsiia; Zionism in Soviet Russia
Joint Distribution Committee (JDC) 5, 8, 9, 11, 12, 50, 51, 52, 60, 64, 70, 73

Kadets *see* Constitutional Democrats
Kadima 67
Kalinin 9, 13
Kaliuzhnyi, I.V. 24
Kamenev, Lev 8, 9, 13, 71
Karakala 111
Kargopol'skii Raion 31
Karlson, Eduard 1; memorandum of 8
Karsavin, Lev 22
Katanian, Ruben 8, 13
Kazakhstan 83
Kazan 121
Kaznachei, Olya 81
Kaznachei, Yehoshu'a (Shunya) 81, 84
kehillot see Jewish communal institutions
Kerki 104
Kfar Bilu 80
Kfar Gil'adi 75
Kfar Oved 80
Kharkov 15, 22, 28, 93, 94, 103, 120
Khiva 120
Khorezm 111
Kiev 6, 7, 13, 15, 90, 120
Kisch, Frederick H. 54, 55, 56, 57, 59, 62, 63, 65, 72, 73, 84, 92, 97, 99, 100, 131
Kishinev 46; pogrom 87
Knesset 81
Kokchetav 111
Komsomol *see* Communist Youth League
Komzet *see* Committee for Agricultural Settlement of Toiling Jews
Kornilova-Serdiukova, L.I. 24
Korolenko, V.G. 28
Kostroma 19
Kostyrchenko, G.V. 12
Kovno 50
Krakow Union for Aid to Prisoners 25
Krasnogorskaia, Esfir 23

Krasnoiarsk 77
Krasnopresnenskaia Prison 18
Krasnovodsk 104, 118
Kremenchug 13, 19, 117
Krestinskii, Nikolai 8, 13
Krimker (Asu'ach), Chaya and Moshe 77
Krym 40
Kuntres 43
Kurgan district 119
Kursk 77, 83, 121
Kurskii, D.I. 21
Kurtamish 119
Kurts, V. 33
Kuskova, Ekaterina 21

Labour Schedule 55, 56, 57, 58, 61, 73, 92, 97, 131
Larin, Iurii 8, 13
Latvia 80, 115
Latvian Zionist Organization 45
League of Nations 41
Left Po'alei Tziyon 67, 73, 126
Legal Hechalutz *see* Hechalutz
Lenin, V.I. 14, 21, 25, 27, 40
Leningrad 6, 15, 19, 28, 30, 57, 62, 120, 123, 127, 128
Lepinsk 123
Levitan, D.M. 31
lishentsy 3, 4, 32
Lithuania 115
Litvinov, Maksim Maksimovich 33, 50, 74
Liubarskii, Solomon 79, 84
Livne (Liberman), Tzvi 68
Look, C.H. 74
Lubianka Prison 14, 25
Lunacharskii, Anatolii 67

Maccabi 15, 75
McDonald, Ramsay 59
Mador, Izrail' 19
Magen, the Society to Help Those Persecuted for Jewishness, Zionism, and All National Matters in Soviet Russia 44, 63, 64, 65, 73
MAPAI 81, 82
Mel'gunov, Sergei 22
Mensheviks *see* Russian Social Democratic Labour Party
Menzhinskii, Viacheslav 15, 17, 30, 39, 91, 94, 95, 130
Mereminsky (Marom), Yisrael 47, 48, 68, 92, 93–4
Merezhin, Avrum 86, 88, 133
Mindel, N.I. 72

Minsk 6, 13, 14, 93, 95
minus zones 18, 34, 77, 79, 83, 84
Mishmar 9, 116, 117
Mishmarot, Kibbutz 80
Molodetskaia-Arav, A.L. 105–11, 132
Moscow: arrests and imprisonment in 7, 15, 17, 18, 19, 22, 31, 95; British mission in 8, 34, 50, 55, 56–7, 62, 64, 71, 96, 97, 98, 109, 127, 128; Jewish aid organizations in 50, 60, 61, 73; Zionist conferences in 7, 14; Zionist organizations in 6, 9, 48, 72, 87, 89, 90
Motzkin, Leo 51

Naamani, N. 71, 72
Nahalal 110
Namir (Nemirovskii), Mordechai 81, 131
Narkomiust 17
Narym 9, 120
Nemirovskii, Mordechai *see* Namir
NEP *see* New Economic Policy
Netherlands 115
New Economic Policy (NEP) 3, 4, 5, 7
New Zealand 114
NKVD *see* People's Commissariat for Internal Affairs
Noel-Baker, Philip 52
Norwegian Legation in Moscow 60, 98, 100, 101
Novonikolaevsk 15
Novorossiisk 22, 40, 54, 67
Nurock, Max 57, 58, 63, 99, 125, 131

Obdorsk 111
October Revolution 2, 4
Odessa 14, 15, 22, 32, 34, 35, 36, 39, 50, 55, 59, 60, 63, 65, 80, 92, 94, 105, 106, 107, 108, 109, 110, 120, 128; arrests of Zionist in 14; British representatives in 56, 72; as transit point for Palestine 23
OGPU (United State Political Directorate) 8, 15, 16, 17, 18, 19, 20, 22, 23, 25, 27, 28, 29, 31, 32, 35, 36, 89, 91, 95, 102, 103, 105; *see also* Cheka; GPU; NKVD; Special Commission of the OGPU
Organization of Russian Zionists 2
OSO *see* Special Commission of the OGPU
Ozet *see* All-Union Society for Agricultural Settlement of Toiling Jews

Pale of Settlement 3, 6, 87
Palestine Government *see* British Mandatory Government of Palestine
Palestine Offices 46, 47, 53, 54, 70, 93, 131; Riga 45; Warsaw 60, 96–7
Palestine Zionist Executive (PZE) 41–9 *passim*, 53, 54, 55, 56, 57–70 *passim*, 72, 92, 93, 97, 99, 101, 102, 122, 123, 131; responsibilities of 131
passport, Soviet 31, 32, 33, 35, 98, 102, 103–4, 103–10 *passim*, 121, 132; fees for 39, 51, 64, 70
Pechenek-Levi, Mania 82
People's Commissariat for Internal Affairs (NKVD) 7, 17, 30, 32, 36, 71
People's Commissariat of Finance (Narkomfin) 32
People's Commissariat of Justice (Narkomiust) 17
People's Will (Narodnaia Volia) 24, 28
Peshkov, Aleksei *see* Gorky, Maxim
Peshkova, Ekaterina 2, 9, 24–35, 47, 60, 61, 64, 65, 73, 78, 97, 98, 99, 100, 101; activities during WWI 26; assistance to Zionists 28, 34–5, 40, 102–10; biography 25–7; on Dzerzhinskii 27; establishment of Pompolit 25, 85–6, 129; and Gorky 25–7, 30; GPU connections 25, 27, 30–1
Petrograd *see* Leningrad
Pick, Chaim 70, 93
Pines, Dan 67, 71, 90, 107
Plumer, Herbert Onslow 58, 61, 72, 73, 100, 102, 131
Po'alei Tziyon 61, 63, 73, 135; *see also* Left Po'alei Tziyon
Podolia 116
pogroms 2, 3, 82, 87, 88, 129; *see also* Kishinev
Poland 22, 24, 26, 27, 33, 80, 95, 115, 126; immigration to Palestine from 57
Polish Red Cross 26
Politburo 17, 24, 86, 88, 96
Political Red Cross *see* Pompolit
politizoliator 17, 115–20; *see also* prisoners and imprisonment
Poltava 28
Poltoratsk 20, 37, 103, 104, 105, 111
Pompolit (Committee to Aid Political Prisoners/Aid to Political Prisoners) 2, 9, 14, 18, 24–35, 37, 41, 43, 47, 58, 60, 65, 97, 98–100, 102–13; demise of 30–1; funding of 24, 29–30, 34; Kharkov office 132; Leningrad office 39; Moscow office 39, 58; *see also* Peshkova

Poslednie novosti 62
Potresov, A.N. 9
Priangarskii region 78
prison and labour camps 17, 76, 115–20
prisoners and imprisonment 14–19, 30, 31, 39, 89–91, 92–3; conditions of 93–4, 97, 117–21; history of 18, 24, 26; hunger strikes 94; *see also* exile, internal; Pompolit
Prokopovich, Sergei 21
PZE *see* Palestine Zionist Executive

Red Army 86, 129
Red Cross 26, 27, 33
Red Cross Society of the People's Will 24
religion, attacks of Soviet state against 3, 5
Remez, David 71
repatriation 26, 38, 127; *see also* Agreement on Repatriation
Riga *see* Palestine Offices
Ritov, Yisrael 97
Romania 80, 115
Rosen, Joseph 8
Rosenblit, Felix 114
Rostov on Don 120
Rozoff, Israel 44
Russian Communist Party (RCP) ix, 6, 12, 33, 73, 86–8, 126
Russian Red Cross 33
Russian Social Democratic Labour Party (Menshevik) 9, 12, 19, 21, 23, 25, 66, 87, 94, 95, 97
Russian Socialist Revolutionary Party (SRs) 12, 25, 26, 66, 87, 97
Russian Society of the Life of the Jews 25
Rykov, Aleksei 8, 13

Sakharov, E.N. 26
Samara 14, 36
Samsonov, T.P. 15
Samuel, Edwin 72
Samuel, Herbert 54, 57, 58, 72, 131
Sandomirskaia, Lota Borisovna 28, 103, 132
Saratov 121
Saunders, A. 122–3
Schechtman, J.B. 12
Schlusselburg Prison 39
Scotland Yard 41, 62, 126, 127
Sevastopol' 40
Shadrinskii Okrug 31
Sheptovitskii, Zalman 77, 83
Shkolnik, Levi *see* Eshkol

Shor, David 8, 9, 22, 71
Shtarkman, Miriam 73
Shvetsov, S.P. 28, 30, 39
Simferopol' 111
Sinegub, L.V. 24
Smena Vekh 87, 88, 129
Smidovich, Petr 8, 9, 13, 71
Smolensk 86
Social Democrats *see* Russian Social Democratic Labour Party
Socialist Revolutionaries *see* Russian Socialist Revolutionary Party
Society for Aid to Political Exiles and Prisoners 24
Solel Boneh 81
Solomon, Harold 72
Solovetskii Labour Camp (Solovki) 17, 18, 93
Solzhenitsyn, Aleksandr 39
Sotsialisticheskii vestnik 89, 130
South Africa 50, 115
Sovnarkom *see* Council of People's Commissars
Sovtorgflot 32, 34, 35
Spaso-Evfimiev Monastery 17
Special Commission of the OGPU (OSO) 17, 18, 22, 23, 24, 29, 31, 32, 95, 135
spetslageria see prison and labour camps
Stalin, Iosif 7, 8, 9, 30, 46, 66, 82
State Political Directorate *see* GPU
Stein, Leonard 59
STP *see* Zionist Toilers' Party
substitution emigration (*zamena*) 1, 12, 13, 20–4, 43, 77, 95; Britain's role in 55–9; decline of 59–64; defined 1; economic considerations for 32, 33; implementation of ix, 23–4, 32–5, 40, 46–55; origins of ix, 9, 10, 21–2; personal narratives of 75–83; and Pompolit 31–5
Surgut 123
Sverdlovsk *politizoliator* 119
Switzerland 115
Symes, George Stewart 58, 101, 131

Taganskaia Prison 18
Taldy-Kurgan 111
Tashkent 77, 103
Tavori (Pipik), Efrayim 81, 84
Tedzhen 104
Tel Chai (training farm) 9, 13, 80, 116
Tiumen' 112, 123
Tobol'sk 13, 17, 40, 45, 108, 111, 112
Tobol'sk *politizoliator* 119

training farms *see* agricultural training farms
Transylvania 115
Troitsk 113
Trotsky, Lev 25
Turkestan 18, 77, 115, 118, 120
Turkey 126, 129; *see also* Constantinople
Turkmeniia 20, 104
Tzap, Kuntsiia 76
Tze'irei Tziyon 5, 9, 12, 15, 22, 68, 89, 135, 136
Tze'irei Tziyon-Hitachdut *see* Zionist Toilers' Party
Tzizling, A. 67
TzS *see* Zionist–Socialist Party
TzS Yugend *see* Zionist Socialist Youth League

Uil 20
Ukraine 1, 3, 4, 6, 8, 15, 19, 25, 56, 86, 120; nationalist organizations in 87
Ul'ianova, Maria 27
Uman' 108
United All-Russian Organization of Zionist Youth (EVOSM) 5, 9, 12, 13, 67, 73, 80, 81, 84, 116, 119, 120, 124, 136
United Committee to Help the Prisoners of Zion (Ha-Va'ad ha-Me'uchad le-Ezrat Shvuyei Tziyon) 44
United States 13, 52, 64; Jews in 50, 51, 52, 69, 114
Unshlikht, Iosif S. 25, 129
Ural district 119
Urals 18, 19, 22, 31, 87, 95, 118, 119, 120
Ust'-Sysol'sk 111

Vdovets, Gonia 79
Verkhne-Ural'sk 17
Verlinsky, Nachum 45, 69
Vest, Binyamin 48, 69, 90, 130
Vienna 46
Vikulovo 119, 123
Vilenchuk, Yitzchak 44, 69
Vinaver, Mikhail L'vovich 28, 29, 30, 31, 47
Vinnitsa 94, 102, 103
visas for Palestine 102, 122, 126; character references for 127; financial guarantees for 127; and Norwegian legation in Moscow 98, 100; procedures for receiving 99

Vitebsk 14, 86
Vladimir 22
Volga region 87
Vologda 79
Voronezh 121
VTsIK *see* All-Russian Executive Committee of Soviets

Warburg, Felix M. 52, 70
War Communism 116
Warsaw 46; British legation in 60; Palestine Office in 60
Weizmann, Chaim 44, 50, 51, 52, 54, 69
West, Benjamin *see* Vest, Binyamin
Wise, Stephen 70
World Hechalutz *see* Hechalutz
World War I 2
World Zionist Organization (London) 2, 41, 50–6, 72, 113, 121
Wrangel, Petr Nikolaevich 88, 129

Yagur 80
Yugend Farband *see* Zionist Socialist Youth League
Yugoslavia 115

zamena see substitution emigration
Ze'ira, Mordechai 81
zemkhozy 88
Zhitomir 14, 36, 123
Zinov'ev Doctrine 45
Zinov'ev telegram 59
Zionism, history of 2; Soviet ideology on and policy toward 3, 12, 14
Zionism in Soviet Russia 1–13; background 2–10; conditions of 113, 115; funding of 40, 50–3, 113–15; movements 42–6; persecution of 6, 7, 12, 15, 17, 24, 36, 89–91, 95, 113, 115, 121; publications 89, 90
Zionist Congresses 44, 45, 51, 113
Zionist Executive *see* Palestine Zionist Executive
Zionist–Socialist Party (TzS) 5, 6, 7, 9, 11, 12, 15, 18, 19, 23, 43, 48, 66, 67, 74, 81, 89, 90, 91, 93, 95, 97, 116
Zionist Socialist Youth League 5, 9, 12, 13, 66
Zionist Toilers' Party (STP, Hitachdut) 5, 6, 7, 11, 12, 15, 43, 44, 52, 67, 73, 74, 84, 89, 90, 117, 123, 136
Znamia truda 26

eBooks – at www.eBookstore.tandf.co.uk

A library at your fingertips!

eBooks are electronic versions of printed books. You can store them on your PC/laptop or browse them online.

They have advantages for anyone needing rapid access to a wide variety of published, copyright information.

eBooks can help your research by enabling you to bookmark chapters, annotate text and use instant searches to find specific words or phrases. Several eBook files would fit on even a small laptop or PDA.

NEW: Save money by eSubscribing: cheap, online access to any eBook for as long as you need it.

Annual subscription packages

We now offer special low-cost bulk subscriptions to packages of eBooks in certain subject areas. These are available to libraries or to individuals.

For more information please contact webmaster.ebooks@tandf.co.uk

We're continually developing the eBook concept, so keep up to date by visiting the website.

www.eBookstore.tandf.co.uk